FLORA OF THE GUIANAS

Edited by

M.J. JANSEN-JACOBS

Series A: Phanerogams
Fascicle 27

71. CYRILLACEAE
(J.C. Yesilyurt)

79. THEOPHRASTACEAE
(B. Ståhl)

86. RHABDODENDRACEAE
90. PROTEACEAE
(G.T. Prance)

100. COMBRETACEAE
(G.A. Stace)

113. DICHAPETALACEAE
(G.T. Prance)

167. LIMNOCHARITACEAE
168. ALISMATACEAE
(R.R. Haynes & L.B. Holm-Nielsen)

including
Wood and Timber
(I. Poole & J. Koek-Noorman)
(F. Lens & S. Jansen)
(J. Koek-Noorman, I. Poole, L.Y.T. Westra & J.W. Maas)

2009
Royal Botanic Gardens, Kew

Contents

71. CYRILLACEAE

by

JOVITA C. YESILYURT[1]

Shrubs or small trees (up to 25 m tall), usually evergreen. Leaves alternate, simple, spirally arranged to usually clustered at stem apex; stipules present or absent, usually inconspicuous or caducous; usually petiolate (rarely sub-sessile); blades glabrous, margin entire, often coriaceous. Inflorescences terminal, sometimes axillary racemes; bracts (single) and bracteoles (paired) present or absent. Flowers bisexual, actinomorphic or slightly zygomorphic, 5-merous, pedicellate; sepals and petals imbricate, free or basally connate, sepals persistent, petals white to pink, usually with inrolled margins; stamens 5 or 10, usually opposite petals, in 1 or 2 (in *Purdiaea* Planch.) whorls, filaments free, anthers 2-locular, opening by apical pores or longitudinal slits, versatile; nectary-disc present around base of ovary or absent; ovary superior, 2-5-locular, style usually short, stigma entire or 1-3(-4)-lobed, ovules 1-3 per locule, pendulous, anatropous. Fruits indehiscent, berries or capsules, dry; seeds 1-few, endosperm copious.

Distribution: A tropical family of 3 genera and ca. 14 species; in the Neotropics only 2 genera occur (*Cliftonia* Banks ex C.F. Gaertn. is basically a N American genus), from south and eastern U.S.A, through Mexico and the West Indies down to northern S America, except the Andes (only *Purdiaea nutans* Planch. has been recorded to reach the Andes); in the Guianas only 1 species.

Note: CYRILLACEAE have traditionally been placed in the ERICALES (sensu Cronquist, 1981), usually close to the ERICACEAE and CLETHRACEAE. These positions have been supported by analyses of DNA sequence data (Anderberg *et al.*, 2002). The family is related to the CLETHRACEAE from which it differs by its indehiscent fruit, number of carpels and locules (3 in *Clethra* L.), presence of nectary disc and few ovules (numerous in *Clethra*). Besides *Cyrilla*, the only other tropical American genus recorded, *Purdiaea* Planch., has its centre of diversification in Cuba, and has been recorded in the Guayana Highlands of Venezuela (Miller, 1998; Ståhl, 2004), Colombia (Thomas, 1960), Ecuador (Ståhl, 1992) and Peru (Pennington *et al.*, 2004).

[1] Herbarium, Royal Botanic Gardens, Kew, Richmond, Surrey, TW9 3AB, U.K.

LITERATURE

Anderberg, A.A. *et al.* 2002. Phylogenetic relationships in the order Ericales s.l.: analyses of molecular data from five genes from the plastid and mitochondrial genomes. Amer. J. Bot. 89: 677-687.

Miller, J.S. 1998. Cyrillaceae. In J.A. Steyermark *et al.*, Flora of the Venezuelan Guayana 4: 663-665.

Pennington, T.D. *et al.* 2004. Illustrated guide to the Trees of Peru. Cyrillaceae. pp. 452-453.

Ståhl, B. 1992. Cyrillaceae. In G. Harling & L. Andersson, Flora of Ecuador 45: 27-34.

Ståhl, B. 2004. Cyrillaceae. In N. Smith *et al.*, Flowering plants of the Neotropics. pp. 125-126.

Thomas, J.L. 1960. A monographic study of the Cyrillaceae. Contr. Gray Herb. 186: 1-114.

KEY TO THE GENERA

1   Flowers actinomorphic, white to pale pink, sepals equal with entire margin; stamens 5, in one whorl, anthers dehiscing by longitudinal slits; style short . . . . *1. Cyrilla*
Flowers zygomorphic, pink to violet, sepals unequal with fimbriate margin; stamens 10, in two whorls, anthers dehiscing by terminal pores; style elongate (to be expected in Guyana) . . . . *Purdiaea*

1. **CYRILLA** Garden ex L., Mant. Pl. 1: 5. 1767.
Type: C. racemiflora L.

For description see under the species.

Distribution: 1 species from southern and eastern U.S.A, through Mexico, the West Indies, C America down to northeast Brazil.

1. **Cyrilla racemiflora** L., Mant. Pl. 50. 1767. Type: A. Garden s.n., Herb. Linnaeus No. 272.1 (lectotype LINN; designated by Nelson, Taxon 39: 665. 1990).   – Fig. 1

Cyrilla brevifolia N.E. Br., Trans. Linn. Soc. London, Bot. ser. 2. 6: 22, t. 1, f. 7-16. 1901. – *C. racemiflora* L. var. *brevifolia* (N.E. Br.) Steyerm., Acta Bot. Venez. 2: 234. 1967. Type: Summit of Mt. Roraima, Im Thurn 334 (lectotype K, isolectotype BM; designated here).

Fig. 1. *Cyrilla racemiflora* L.: A, habit; B, portion of the inflorescence, showing the bract at the base of pedicel; C, open flower, showing the 2 bracteoles, sepals, petals, stamens and gynoecium; D, fruit, with persistent bracteoles and sepals; E, bract; F, petal; G, sepal; H-M, leaf variation (A-C, G, Maguire & Fanshawe 23106; D-F, M, Hoffman *et al.* 1870; H, Maguire *et al.* 46025A; J, Tillet *et al.* 45102; K, McDowell & Gopaul 2831; Pipoly & Gharbarran 9516). [Scale bars: single bar scale (–) represents 1 mm, double bar (=) 1 cm]. Drawing by Linda Gurr.

4

Shrub or small tree, up to 7-15 m tall, sometimes multistemmed, occasionally spreading with vine-like branches, rarely branched at base. Leaf scars prominent. Stipules inconspicuous (caducous) or absent; petiole usually short (nearly subsessile), slightly winged, 0.4-1.3 cm long; blade subcoriaceous to coriaceous, oblanceolate to elliptic, occasionally linear, 2-11(-12) (including petiole) x 0.4-2.5(-3) cm, margin usually slightly reflexed, apex acute to rounded, sometimes emarginated (rarely apiculate), base narrowly cuneate to attenuate; midvein generally depressed above, prominent beneath, densely reticulate, slightly prominent or obscure above, secondary veins straight to slightly arcuate, divergent, irregularly anastomosing. Inflorescence racemose, axillary, usually near end of branches, glabrous; peduncle 0.4-1.7 cm long (before the first bract); rachis up to 14 cm long, usually somewhat ridged; often numerous flowers, erect or pendulous; bract lanceolate-subulate, 0.8-2.4(-3.5) mm long, acuminate to very acute, basally thickened and usually costate, persistent; bracteoles 2, usually above middle of pedicel, alternate or opposite, lanceolate, 1-1.5(-1.6) mm long, persistent; pedicels ascending or slightly patent, alternate or subopposite, 1.1-2 mm long (up to 4 mm when fruiting), usually pink. Flowers actinomorphic; sepals free, or basally connate, 5, pink, coriaceous, usually with basal thickening, lanceolate-ovate to deltoid, 0.6-1.5(-1.6) x up to 0.7(-1) mm, apex acute to acuminate, glabrous, persistent; petals imbricate or slightly contorted, free or shortly fused, 5, white, usually more membranaceous along margin, oblong to oblong-lanceolate or oblong-elliptic, 2(-2.4) x up to 1.1 mm, margins sometimes inrolled (giving the appearance of long acuminate), apex acute, obtuse or rounded, glabrous to rarely glandular on lower inner surface; stamens 5, about 2/3 as long as petals (always shorter), filaments flattened, 1.1-1.4 mm long, anthers terete, linear-subulate, up to 0.4 mm broad, very occasionally with mucronate tip, versatile, dehiscing longitudinally, pink or purple; nectary disc sometimes present; ovary 2-3(rarely-4)-locular, oblong to slightly ovate, wrinkled when dried, glabrous, up to 1.6(-2) mm long, ovules usually 2, pendulous from apex of locule, style 1, short, stout, stigma entire, or 2-3-lobed, corresponding to number of locules. Fruit dry, indehiscent, green immature, red (pink) at maturity, subglobose to globose or conic-ellipsoid, 2.5-(3) x up to (0.9)2(3) mm, usually smooth (rarely slightly rugulose), glabrous, stigma usually persistent; seeds 1 per locule, elongate.

Distribution and ecology: SE U.S.A., Mexico, C America, West Indies, Venezuela, the Guianas and Brazil; in several different habitats including savannas, gallery forest, swampy vegetation and cloud forest, usually on white sands or along river banks, elev. 25 to 2300 m; mentioned often as fire resistant; flowers sometimes with galls; 226 specimens examined of which 97 from the Guianas (GU: 93; SU: 3; FG: 1).

Selected specimens: Guyana: SSE Georgetown, vicinity of St. Cuthbert's Village, Mori *et al.* 8044 (K, NY, U); Rockstone, Gleason 486 (GH, K, NY, US); Kaieteur Plateau, Maguire *et al.* 23106 (K, BM, US); Potaro-Siparuni Region, Kaieteur Nat. Park, Gillespie *et al.* 917 (CAY, K, NY, U, US); Upper Mazaruni R., Arubaru R., Kako tributary, Pinkus 190 (GH, NY, US); Potaro-Siparuni Region, Pakaraima Mts., Mutchnick *et al.* 182 (CAY); E Berbice-Corentyne Region, Canje R., W of Digitima Cr., Pipoly *et al.* 9516 (B, NY, U, US); .Mazaruni-Potaro Region, near Lake Gladys, Liesner 23327 (VEN). Suriname: Maratakka R., Boschwezen = BW 946, 3429 (U, NY); Maratakka R., Saparra Cr., Maas, LBB 10868 (NY, U). French Guiana: Oyapock R., Oldeman B-1474 (CAY).

Uses: Planted as an ornamental in the USA. In the West Indies, the species is esteemed among honey producers due to its nectar-rich flowers, which are present over an extended period of time (Ståhl, 2004).

Vernacular names: Guyana: warimiri howadanni, kwako, kakiratti (Arawak), krawasjirang, karawasjipio (Karib), koewaliroemang (Arawak).

Notes: The examined specimens showed a striking and considerable amount of variation in size and shape and even texture in almost all the features from the leaves, flowers and fruits. It is difficult to assess whether one or more of these characters might characterise distinguishable taxa. Despite Thomas (1960) acknowledges that some characters (e.g. leaves, mucronate stamens, bracts and fruits) exhibit interesting variation patterns, the author concluded that *Cyrilla racemiflora* is very polymorphic and should be considered as a single species. For the same reason, all these variable phenotypes have been considered part of *Cyrilla racemiflora* for the region studied.

The collection Im Thurn 334, is mounted on the same sheet with Quelch & McConnell 88 and 318. Im Thurn 334 is chosen as lectotype of *Cyrilla brevifolia* N.E. Brown, it has branches with flowers and branches with fruits. Moreover, there is a duplicate housed at BM.

EXTRA LIMITAL TAXA

PURDIAEA Planch., London J. Bot. 5: 250. 1846.
Type: P. nutans Planch.

Distribution: Ca. 12 species in C America, Cuba, Colombia, Venezuela, Ecuador, Peru, most of them occurring in Cuba, 1 species in the Guayana Highlands of Venezuela.

Purdiaea nutans Planch., London J. Bot. 5: 251. 1846. Type: Colombia, New Granada, La Cruz, Purdie s.n. (K).

Evergreen trees or shrubs. Leaves usually crowded at branch tips, generally sessile. Inflorescence generally terminal, racemose, bracts scarious, bracteoles absent; sepals unequal (outer sepals lobes conspicuously enlarged and with ciliate/fimbriate margins, often enlarged in fruit; stamens 10, in two whorls, anthers opening by apical pores; ovary usually 5-locular, 1-ovulate, style elongate. Fruit indehiscent, 3-5-ribbed, dry, enclosed in persistent sepals, seeds 1-5.

N o t e : *Purdiaea nutans* has been recorded for the Guayana Highlands (Miller, 1998; Ståhl, 2004), usually in upper montane (tepui) forests and cloud forest.

79. THEOPHRASTACEAE

by

BERTIL STÅHL[2]

Shrubs or small trees, usually evergreen. Leaves alternate, often pseudoverticillate, exstipulate, petiolate, simple, glandular-punctate, mostly with bundles or layers of subepidermal, extraxylary sclerenchyma. Inflorescences terminal or lateral, racemose, rarely reduced to a single flower, each flower subtended by a small bract. Flowers regular, or because of unequal size of the corolla lobes, slightly zygomorphic, 5- or sometimes 4-merous, bisexual or (usually in *Clavija*) unisexual, aestivation imbricate; calyx persistent, lobes free to base, glandular-punctate, margins membranaceous; corolla sympetalous, usually firm and waxy, lobes usually slightly unequal in size; staminodes fused to corolla tube, alternating with lobes; stamens of same number as corolla lobes and placed in front of these, filaments flattened, fused to lower part of corolla, free to base, or (often in *Clavija*) united into a tube, anthers basifixed, dithecal, extrorsely dehiscent with longitudinal slits, thecae partly filled up with calcium oxalate crystals; pistil superior, ovary ovoid to subglobose, undivided, 1-locular, ovules few to numerous, spirally inserted on a basal column, style often poorly demarcated, shorter to about as long as ovary, stigma truncate, capitate or sometimes discoid, entire or vaguely lobed. Fruits berries with dry and sometimes woody pericarp, indehiscent, subglobose, oblongoid, or ovoid, yellow, orange, or orange-red; seeds embedded in juicy and (when mature) sweet pulp, hard, brown to brownish-yellow, endosperm abundant.

Distribution: Throughout the Neotropics but most diverse in and around the Antilles; ca. 100 species in 7 genera; 1 genus with 3 species in the Guianas.

LITERATURE

Källersjö, M. & B. Ståhl. 2003. Phylogeny of Theophrastaceae. Int. J. Pl. Sci. 164: 579-591.

Lindeman, J.C. 1979. Theophrastaceae. In A.L. Stoffers & J.C. Lindeman, Flora of Suriname 5(1): 367-369.

Miller, J.S. & B. Ståhl. 2005. Theophrastaceae. In J.A. Steyermark *et al.*, Flora of the Venezuelan Guayana 9: 325-329.

---

[2] Gotland University, SE 62167 Visby, Sweden.

Ståhl, B. 1991. A revision of Clavija (Theophrastaceae). Opera Bot. 107: 1-77.

Ståhl, B. 2002. Theophrastaceae. In S.A. Mori *et al.*, Guide to The Vascular Plants of Central French Guiana. Part 2. Mem. New York Bot. Gard. 76(2): 706-709.

1. **CLAVIJA** Ruiz & Pav., Fl. Peruv. Prodr. 142, t. 30. 1794.
   Type: C. macrocarpa Ruiz & Pav.

Shrubs or small trees, unbranched or sparsely branched, evergreen. Leaves pseudoverticillate; short- to long-petiolate; blades medium-sized to large or very large, margins entire, serrulate or serrate, sometimes spinose-serrate; extraxylary sclerenchyma present or sometimes lacking. Inflorescences lateral racemes, solitary or in small groups, borne on stems among and beneath leaves, few- to many-flowered, those of female plants usually shorter than those of male plants; bracts lanceolate-ovate, inserted at junction of pedicel and rachis, rarely ascending on pedicels. Flowers 5- or sometimes 4-merous, uni- or bisexual, dioecious, gynodioecious, androdioecious, or rarely hermaphroditic; calyx greenish, lobes suborbicular, margin erose; corolla pale to dark orange, crateriform, not glandular-punctate, lobes broadly ovate-oblong or suborbicular; staminodes oblongoid to ovoid, inserted at mouth of corolla tube; stamens in non-female flowers united to a tube, in female flowers separated, filaments glabrous, anthers ovoid, obtuse-apiculate at apex; ovary in female flowers ovoid or broadly ovoid, in male or functionally male flowers narrowly ovoid to linear, ovules usually few, style shorter than to as long as ovary, stigma truncate. Fruits yellow, subglobose, pericarp thin and brittle when dried, or sometimes woody; seeds large, obtuse-angled, light to dark brown.

Distribution: About 55 species, from Nicaragua to S Brazil and N Paraguay, 1 species in Haiti and 3 in the Guianas.

KEY TO THE SPECIES AND INFRASPECIFIC TAXA

1   Leaf blade to 35(-40) cm long with serrate-spinose margin; petiole to 3.5 cm
    long . . . . *3. C. macrophylla*
    Leaf blade to 75(-80) cm long with entire, serrulate or sometimes sparsely
    serrate margin; petiole to 8 cm long 2

2 Leaf blade subherbaceous, surface between veinlets smooth or very sparsely striate (extraxylary sclerenchyma lacking or very sparse . . . *2. C. imatacae* Leaf blade coriaceous to subcoriaceous, surface between veinlets striate, at least beneath 3

3 Young shoots glabrous or subglabrous; petiole poorly demarcated (leaf blade narrowly attenuate at base) . . . . . *2a. C. lancifolia* subsp. *lancifolia* Young shoots puberulous; petiole well demarcated (leaf blade attenuate to short-attenuate at base) . . . . *2b. C. lancifolia* subsp. *chermontiana*

1. **Clavija imatacae** B. Ståhl, Opera Bot. 107: 62. 1991. Type: Venezuela, Bolívar, Liesner & González 11105 (holotype MO, isotype GB).

Shrub or small tree to 4 m high; young shoots glabrous or sparsely puberulous at apex. Petiole 1-6.5 cm long, 1.8-3 mm thick, glabrous or sparsely puberulous at base; blade subherbaceous, oblanceolate, 23-76 x 4.5-17 cm, apex acute or short-acuminate, base attenuate, margin serrulate, sometimes subentire or sparsely serrate, often repand, glabrous or sometimes sparsely glandular-pilose beneath; lateral veins and veinlets conspicuous, surface between veinlets smooth or sometimes very sparsely striate above. Inflorescence in bisexual and male plants to 17 cm long with 20-40 flowers, in female plants to 5.5 cm long with up to ca. 22 flowers; rachis 0.8-2 mm long, glabrous or sparsely glandular-puberulous; pedicels 1.5-4 mm long; bracts 0.5-1.5 mm long. Flowers 5-merous; calyx lobes very broadly ovate, 1.7-2.2 x 2-2.5 mm, glabrous; corolla orange, tube 1.7-2 mm, lobes 3-4.5 x 3.2-5 mm; stamen filaments in male and bisexual flowers fused into a tube 0.5-1.2 mm long, in female flowers free, ca. 0.7 mm long; pistil in male and bisexual flowers narrowly ovoid, in bisexual flowers with 2-10 ovules, in female flowers ovoid with ca. 12 ovules. Fruit orange, 1.8-2.5 cm diam., pericarp ca. 0.3 mm thick, outside smooth; seeds 2-8, 5.5-8.5 mm long.

Distribution: NW Venezuela and adjacent Guyana; primary forest, from sea level to 350 m elev.; 20 collections studied (GU: 2).

Specimens examined: Guyana: Cuyuní R., below Akaio Falls, Sandwith 635 (K, NY, U); Takutu Cr. to Puruní R., Mazaruni R., Fanshawe 2058 (K, NY, U).

Phenology: Flowering ± continuously throughout the year.

2. **Clavija lancifolia** Desf., Nouv. Ann. Mus. Hist. Nat. 1: 402. 1832. Type: [icon] Desf., Nouv. Ann. Mus. Hist. Nat. 1: t. 14 (lectotype).

Shrub to 3 m, sometimes higher; young shoots puberulous or subglabrous. Petiole 1-8 cm long, 1-3 mm thick, puberulous at base, or glabrous; blade coriaceous or subcoriaceous, oblanceolate, narrowly oblanceolate, or sometimes narrowly ovate, obovate or elliptic, 12-60 x 3.5-14 cm, apex acute or acuminate, base attenuate, margin entire, serrulate or rarely sparsely serrate, glabrous or rarely glandular pilose beneath; lateral veins rather inconspicuous, at least above, surface between veinlets striate, furrowed or sometimes smooth above, striate beneath. Inflorescence in male and bisexial plants to 20 cm long with (8-)15-40 flowers, in female plants to 4 cm long with 5-12 flowers; rachis 0.5-1.4 mm thick, puberulous or sparsely puberulous; pedicels 1-3 mm long; bracts 0.5-1.2 mm long. Flowers 5-merous; calyx lobes very broadly ovate, 1.5-2.5 x 1.5-3 mm, glabrous; corolla orange or orange-red, tube 1.5-3 mm long, lobes 3-4(-5) x 3.2-5 mm; stamen filaments in male and bisexual flowers fused into a tube 0.7-1.5 mm long, in female flowers free, 0.5-1 mm long; pistil in male flowers linear or rudimentary, in bisexual flowers sublinear with 1-5(-8) ovules in a single row, in female flowers ovoid with (7-)12-18 ovules in 2 or 3 rows. Fruit orange or orange-yellow, 1-3 cm diam., pericarp 0.2-0.4 mm thick, outside smooth; seeds 2-12, (5-)7.5-10(-13) mm long.

Distribution: Throughout most of the Amazon basin except the western parts.

2a. **Clavija lancifolia** Desf. subsp. **lancifolia**

Clavija ornata D. Don var. *subintegra* A. DC., Prodr. 8: 148. 1844. Type: French Guiana, Cayenne, Poiteau s.n. (lectotype G, isolectotypes FI, P, W).

Young shoots sparsely puberulous to subglabrous. Petiole 1-5.5 cm long, 1.2-2.5 mm thick, glabrous; blade subcoriaceous, oblanceolate or narrowly oblanceolate, 17-40(-60) x 4.5-8.5(-14) cm, apex acuminate or acute, base narrowly attenuate, margin serrulate, glabrous, surface between veinlets striate on both sides but sometimes sparsely so beneath.

Distribution: N Suriname, central and coastal French Guiana, and NE Brazil; primary forest, from sea level to 350 m elev.; 32 collections studies, the majority from the Guianas (SU: 7; FG: 24).

Selected specimens: Suriname: Suriname R., below Kabel, near Jandé Cr., Lindeman 4463 (NY, S, U); Brownsberg Forest Reserve, BW 3857 (U, US). French Guiana: Grand Inini R., between St. Emérillon and Crique Saï, de Granville 657 (CAY, P); Ile de Cayenne, de Granville 5581 (BR); Dégrad Claude, Tamouri, Lescure 183 (CAY); Chemin Maripa, Oyapock, Oldeman T-806 (CAY, GB, P); St. Pararé on Arataye R., Oldeman 2905 (CAY).

Phenology: Flowering mainly in March through August.

2b. **Clavija lancifolia** Desf. subsp. **chermontiana** (Standl.) B. Ståhl, Opera Bot. 107: 56. 1991. – *C. chermontiana* Standl., Publ. Field Mus. Nat. Hist., Bot. Ser. 8: 148. 1930. Type: Brazil, Pará, Dahlgren & Sella 200 (holotype F). – Fig. 2

Young shoots densely puberulous. Petiole 1.5-8 cm long,1-3 mm thick, puberulous at base; blade coriaceous, oblanceolate, sometimes narrowly obovate or elliptic, 14-45(-53) x 3.5-12(-14) cm, apex acute or short-acuminate, base attenuate, margin entire, serrulate, or rarely sparsely serrate, glabrous or sometimes glandular-pilose beneath, surface between veinlets striate, furrowed or sometimes smooth above, striate below.

Distribution: Central and S Venezuela, the Guianas, E and central Amazon, and N Bolivia; primary forest, from sea level to 750 m elev.; more than 250 collections studied, about 45 from the Guianas (GU: 5; SU: 28; FG: 11).

Selected specimens: Guyana: Kanuku Mts., trail to Mt. Iraimakipang, Goodland & Maycock 462 (US); foothills of NW Kanuku Mts., near Moco-Moco, Maas *et al.* 3786 (F, K, NY, P, U); W Kanuku Mts., drainage of Takutu R., A.C. Smith 3218 (A, B, F, G, LIL, MO, NY, P, S, U, US, W). Suriname: Raleigh Falls, Coppename R., Boon 1013 (U); Corantijn R., Matappi, BW 2173 (U); Emmaketen, below Grote Hendriktop, Daniëls & Jonker 991 (U); 2 km S of Juliana Top, 13 km N of Lucie R., Irwin *et al.* 54617 (GH, MG, NY, P, S, U, US). French Guiana: Tortue Mts., Itany (Haut Maroni), Cremers 4680 (CAY); Saül, La Fumée Mts., Mori *et al.* 17928 (GB); Antecum-Pata on Itany R., Sastre 1409 (CAY).

Vernacular names: Suriname: aimiara enoeroe, wintje-bobi.

Phenology: Flowering mainly in May through September.

12

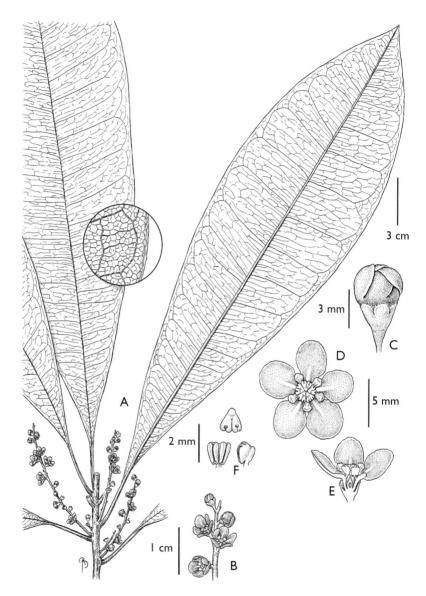

Fig. 2. *Clavija lancifolia* Desf. subsp. *chermontiana* (Standl.) B. Ståhl: A, flowering shoot; B, tip of raceme; C, bud just prior to anthesis; D, flower seen from above showing petals, gibbous staminodes, and anthers (male or morfologically bisexual flower); E, flower in longitudinal section; F, anther (dorsal, abaxial and lateral view). Drawing by B. Angell; reproduced with permission from S.A. Mori *et al.*, Guide to The Vascular Plants of Central French Guiana. Part 2. Mem. New York Bot. Gard. 76(2): 710, fig. 305. 2002.

3. **Clavija macrophylla** (Link ex Roem. & Schult.) Miq., Fl. Bras.
10: 275. 1856. − *Theophrasta macrophylla* Link ex Roem. &
Schult., Syst. Veg. 4: 187. 1819. Type: Brazil, Pará, Sieber s.n.
(lectotype BR).

Shrub to at least 2 m high; young shoots glabrous or subglabrous. Petiole
0.5-3.5 cm long, 1.3-2.4 mm thick, glabrous; blade coriaceous,
oblanceolate or sometimes narrowly obovate, 16-37 x 4-13 cm, apex
acute or short-acuminate, base narrowly attenuate or attenuate, margin
spinose-serrate, with 15-35 teeth per side, glabrous; lateral veins rather
inconspicuous, surface between veinlets striate or sometimes smooth.
Inflorescence in male and bisexual plants to 14 cm long with 5-35
flowers, in female plants to ca. 3.5 cm long with up to ca. 15 flowers;
rachis 0.5-1 mm thick, puberulous or subglabrous; pedicels 1-2 mm
long; bracts 0.7-1 mm long, inserted at nodes. Flowers 5-merous; calyx
lobes very broadly ovate, 1.5-2 x 1.8-2.2 mm long, glabrous; corolla
orange, tube 2-2.2 mm long, lobes 3-4 x 3-4.2 mm; stamen filaments in
male and bisexual flowers fused into a tube 0.5-1 mm long, in female
flowers free, ca. 0.5 mm long; pistil in male flowers sublinear, in
bisexual flowers narrowly ovoid with 4-20 ovules, in pistillate flowers
ovoid with 15-25 ovules. Fruit orange-yellow, 1-3 cm diam., pericarp
0.2-0.4 mm thick, outside smooth; seeds 1-9, 5.5-11 mm long.

Distribution: SW Guianas and NE Brazil; in secondary dry forest,
from sea level to 600 m elev.; 30 collections studied, 8 of which from the
Guianas (GU: 9).

Selected specimens: Guyana: Takutu R., W parts of Kanuku Mts.,
A.C. Smith 3172 (A, NY); Rupununi R., near mouth of Chairwair Cr.,
A.C. Smith 2365 (A, F, K, NY, US); Rupunini Distr., Shea Rock, Jansen-
Jacobs et al. 4845 (GB, U); Pirara, Ro. Schomburgk ser II, 419 (BM,
CGE, F, G, K, OXF, P, U, W).

Phenology: Flowering from December through May.

86. RHABDODENDRACEAE

by

GHILLEAN T. PRANCE[3]

Shrubs or small trees. Leaves entire, alternate, gland-dotted, coriaceous, small peltate hairs on undersurface; stipules small, subulate or obscure. Inflorescences of supra-axillary racemose panicles or racemes; bracts and bracteoles small, reduced to scales. Flowers hermaphrodite; receptacle broad, slightly concave; calyx very short, lobes 5 or indistinct; petals 5, caducous, sepaloid, oblong or oblong-elliptic, apex rounded or minutely apiculate, minutely punctate, aestivation imbricate; disk absent; stamens numerous (about 45), filaments short, flattened, persisting after flowering and then recurved, anthers linear, erect, basifixed, caducous, 4-locular, dehiscing longitudinally; ovary sessile, globose, glabrous, 1-locular, inserted at base of concave receptacle, ovule 1, basally attached, campylotropous, style arising from base of ovary to one side of it, fairly thick, elongated, stigmatic surface on outermost side ascending from base or middle. Fruits small drupes, globose, terminating a short stipe in cup-shaped receptacle, exocarp thin, crustaceous when dry, endocarp slightly woody; seed 1, reniform-globose, exalbuminous, with thin testa; cotyledons thickly fleshy, radicle small and bent inward towards hilum.

Distribution: The family consist of 1 genus with 3 species distributed in the Guianas, northern Amazonian and NE Brazil; in the Guianas 1 species.

Systematic position: The genus *Rhabdodendron* has been placed variously in the RUTACEAE (Puff & Weber, 1976 and others), CHRYSOBALANACEAE and near the PHYTOLACCACEAE (Prance 1968). The taxonomic history has been described in Prance (1968, 1972). More recent molecular work has confirmed its proximity to PHYTOLACCACEAE as a member of the Caryophyllid group.

LITERATURE

Engler, H.G.A. 1931. Rutaceae subfam. Rhabdodendroideae. In H.G.A. Engler & K.A.E. Prantl, Die natürlichen Pflanzenfamilien ed. 2. 19a: 213, 357-358.
Fay, M.F., et al. 1997. Familial relationships of Rhabdodendron (Rhabdodendraceae): Plastid rbcL sequences indicate a Caryophyllid placement. Kew Bull. 52: 923-931.

---

[3] Royal Botanic Gardens, Kew, Richmond, Surrey, TW9 3AB, U.K.

Huber, J. 1909. Rhabdodendron. Bol. Mus. Goeldi Hist. Nat. Ethnogr. 5: 425-431.

Prance, G.T. 1968. The systematic position of Rhabdodendron Gilg & Pilg.. Bull. Jard. Bot. Natl. Belg. 38: 127-146.

Prance, G.T. 1972. Rhabdodendraceae. Flora Neotropica Monograph 11: 1-22.

Puff, C. & Weber, A. 1976. Contributions to the morphology, anatomy, and karyology of Rhabdodendron, and a reconsideration of the systematic position of the Rhabdodendronaceae. Pl. Syst. Evol. 125: 195-222.

1. **RHABDODENDRON** Gilg & Pilg., Verh. Bot. Vereins Prov. Brandenburg 47: 152. 1905.

Type: R. columnare Gilg & Pilg. [= R. macrophyllum (Spruce ex Benth.) Huber]

Description as for family.

1. **Rhabdodendron amazonicum** (Spruce ex Benth.) Huber, Bol. Mus. Goeldi Hist. Nat. Ethnogr. 5: 427. 1909. – *Lecostemon amazonicum* Spruce ex Benth., Hooker's J. Bot. Kew Gard. Misc. 5: 295. 1853. Type: Brazil, Pará, Santarém, Spruce 377 (holotype K, isotypes LD, MG, OXF, P). – Fig. 3

Lecostemom crassipes Spruce ex Benth., J. Bot. Kew Gard. Misc. 5: 295. 1853. – *Rhabdodendron crassipes* (Spruce ex Benth.) Huber, Bol. Mus. Goeldi Hist. Nat. Ethnogr. 5: 428. 1909. Type: Brazil, Amazonas, Manaus, Spruce 1497 (holotype K, isotypes BM, CGE, M, NY, OXF, P).
Lecostemon crassipes Spruce ex Benth. var. *cayennense* Benth., Hooker's J. Bot. Kew Gard. Misc. 5: 296. 1853. Type: French Guiana, Martin s.n. (holotype K, isotype BM).
Rhabdodendron duckei Huber, Bol. Mus. Goeldi Hist. Nat. Ethnogr. 5: 428. 1909. Type: Brazil, Pará, Faro, Ducke MG 8546 (holotype MG, isotypes RB 8559, US).
Rhabdodendron paniculatum Huber, Bol. Mus. Goeldi Hist. Nat. Ethnogr. 5: 429. 1909. Type: Brazil, Pará, Obidos, Ducke MG 8854 (holotype MG, isotypes BM, US).
Rhabdodendron longifolium Huber, Bol. Mus. Goeldi Hist. Nat. Ethnogr. 5: 430. 1909. Type: Brazil, Pará, Faro, Ducke MG 8504 (lectotype MG, isolectotype RB 8559).
Rhabdodendron arirambae Huber, Bol. Mus. Goeldi Hist. Nat. Ethnogr. 5: 430. 1909. Type: Brazil, Pará, Alto Ariramba, Ducke MG 8000 (not seen).
Lecostemon sylvestre Gleason, Bull. Torrey Bot. Club 54: 608. 1927. – *Rhabdodendron sylvestre* (Gleason) Maguire in Maguire *et al.*, Bull. Torrey Bot. Club 75: 397. 1948. Type: Guyana, between Kangaruma and Potaro Landing, Gleason 211 (holotype NY, isotype GH).

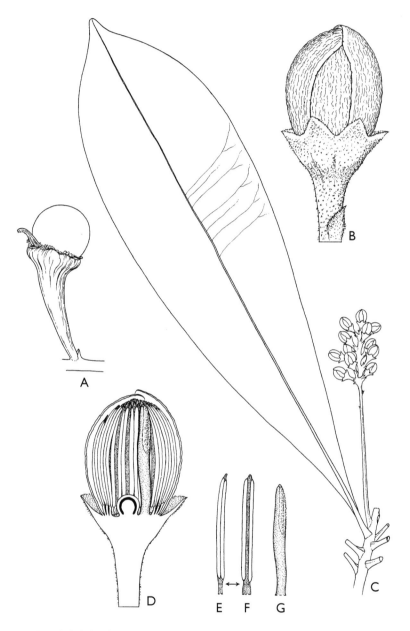

Fig. 3. *Rhabdodendron amazonicum* (Spruce ex Benth.) Huber: A, young fruit (x 0.5); B, flower bud (x 5); C, habit (x 0.5); D, flower section (x 5); E, stamen, lateral view (x 5); F, stamen, dorsal view (x 5); G, style (x 5). Drawing by Charles C. Clare; adapted with permission from Flora Neotropica Monograph 11.

Tree up to 15 m x 20 cm diam., usually smaller; stem with thin hard bark, wood with anomalous secondary phloem; young branches with scattered peltate hairs. Stipules absent. Petiole 1.5–3.5 cm long, with scattered peltate hairs, not winged, terete; blade coriaceous, oblanceolate, oblong to oblong-obovate, 2-39 x 3-10 cm, apex acute, acuminate or mucronate, most frequently with acumen 2-9 mm long, gradually narrowed to cuneate base, glabrous above, with few scattered peltate hairs beneath, not rugose on surfaces; midvein plane to prominulous above, prominulous beneath, secondary veins 30-45 pairs, plane to prominulous above, prominulous beneath, anastomosing but not forming conspicuous marginal vein. Inflorescence supra-axillary and sometimes terminal panicles or occasionally reduced to racemes, 9-17 cm long, sparsely peltate pubescent becoming glabrous with age; bracts and bracteoles ovate to lanceolate, 1-2 mm long, persistent, chartaceous; pedicels 6-15 mm long, glabrescent, frequently recurved, often with 2 lanceolate bracteoles; calyx-tube turbinate-campanulate, 2-4 mm long, exterior glabrescent, lobes small but distinct and apparent in young flowers only; petals 5, oblong, 7-8 mm long, sepaloid, minutely punctate; stamens ca. 45, anthers ca. 7 mm long; stigmatic surface long and linear. Fruit subglobose, 6-10 mm diam., exocarp glabrous, smooth but wrinkled when dry, mesocarp very thin, fleshy, endocarp thin, bony, fragile, with median line of fracture, glabrous within.

Distribution: The Guianas and central to eastern Amazonian Brazil; forest on non-flooded sandy ground; many collections studied (GU: 15; SU: 7; FG: 5).

Selected specimens: Guyana: Demerara-Berbice Region, Fairview Landing, E bank of Essequibo R., McDowell 3251 (U, US); Upper Demerara Region, Mabura Hill, ter Steege et al. 399 (U). Suriname: Tumac Humac Mts., Acevedo-Rodriguez 6067 (U, US); Lely Mts., SW plateau, Lindeman & Stoffers et al. 785 (U). French Guiana: E of Pic du Grand Croissant, 40 km N of Camopi, Feuillet 1209 (CAY, U); Route de Cayenne, km 8.5, Serv. Forestier 7406 (U).

Phenology: Flowering June to October in the Guianas.

Vernacular names: Guyana: mabua (Arawak). Suriname: powisitere.

Note: Lemée, Fl. Guyane Française (2: 186. 1952) reports R. macrophyllum (Spruce ex Benth.) Huber to have been found in French Guiana. So far this species, that is close to R. amazonicum, is known only from the vicinity of Manaus, Brazil.

18

90. PROTEACEAE

by

GHILLEAN T. PRANCE[4]

Trees or shrubs. Leaves alternate, opposite or whorled, simple, pinnatifid, pinnate or bipinnate, usually coriaceous, exstipulate. Inflorescences simple or compound, axillary or terminal, racemose or paniculate. Flowers usually actinomorphic, bisexual, solitary or paired in axils of bracts, rarely ebracteate; tepals 4 valvate, free or variously united, each with a slightly expanded limb; stamens 4, usually all fertile, opposite tepals, filaments partly or wholly adnate to tepals, rarely free; hypogynous glands usually present, 4, scale-like or fleshy, free or fused; pistil 1, ovary superior, rarely perigynous, sessile or stipitate, 1-locular, ovules 1-many, variously inserted, style simple often persistent, often with apex expanded as a pollen presenter, stigma small, terminal or subterminal. Fruits dehiscent or indehiscent, woody or coriaceous follicles, drupes or achenes; seeds 1-many, usually endospermic.

Distribution: A largely southern hemisphere family consisting of 79 genera and ca. 1700 species with Australia and southern Africa as its centers of greatest diversity; 8 genera and 84 species occur in the Americas from Mexico to Chile and Argentina, with the greatest diversity in the Andes and eastern Brazil, some American species also occur in tropical Africa, Madagascar, India, eastern Asia, Malesia, New Caledonia, New Zealand and Fiji; 3 genera and 9 species are known to occur in the Guianas.

LITERATURE

Mennega, A.M.W. 1934. Proteaceae. In A.A. Pulle, Flora of Suriname 1(1): 154-157.
Mennega, A.M.W. 1968. Proteaceae. In A.A. Pulle & J. Lanjouw, Flora of Suriname, Additions and corrections 1(2): 315-320.
Plana, V. 2002. Proteaceae. In S.A. Mori et al., Guide to the Vascular Plants of Central French Guiana. Mem. New York Bot. Gard. 76(2): 592-594.

[4] Royal Botanic Gardens, Kew, Richmond, Surrey, TW9 3AB, U.K.

Acknowledgment. I thank Katie Edwards and Vanessa Plana who collaborated with the monograph of the genera of Proteaceae on which this account is based, and the Royal Botanic Gardens, Kew for facilities to carry out this work.

Prance, G.T. *et al*. 2007. Proteaceae. Flora Neotropica 100: 1-216.
Sleumer, H. 1954. Proteaceae americanae. Bot. Jahrb. Syst. 76: 139-211.
Steyermark, J.A. 2004. Proteaceae. In J.A. Steyermark *et al*., Flora of the
Venezuelan Guayana 8: 384-393.

KEY TO THE GENERA

1　Adult leaves pinnate 2
　　Adult leaves entire, simple or pinnatifid, never pinnate; fruit indehiscent　. . 3

2　Fruit indehiscent, seeds not winged; style recurved . . . . *1. Euplassa*
　　Fruit a follicle, seeds winged; style erect . . . . *3. Roupala*

3　Fruit indehiscent with thick hard or fleshy pericarp, seeds not winged . . . .
　　. . . . . *2. Panopsis*
　　Fruit a dehiscent follicle, pericarp thin not fleshy, seeds winged . . . .
　　. . . . . *3. Roupala*

1.　**EUPLASSA** Salisb. ex Knight, Cult. Prot. 101. 1809.
Type: E. meridionalis Knight, nom. illeg. (Roupala pinnata Lam.,
Euplassa pinnata (Lam.) I.M. Johnst.)

Trees, less frequently shrubs. Leaves spirally arranged, paripinnate, rhachis frequently ending in a terminal appendix extending up to 2 cm beyond terminal leaflet pair, rarely forming a terminal leaflet; leaflets 2-10 pairs, opposite to strongly subopposite, subsessile to long-petiolate, shape various, basal pair consistently smaller than rest; margin entire to remotely serrate. Inflorescences generally unbranched, pseudo-racemose, axillary or rarely terminal, solitary, occasionally with 2 inflorescences per leaf axil; flowers in pairs, each pair supported by a peduncle of variable length which is frequently altogether absent; each pair subtended by a small common bract; flowers sessile or pedicellate. Flowers weakly zygomorphic; tepals all recurving at anthesis or one (innermost) remaining erect; stamens epitepalous, anthers ovate, subsessile, housed in concave distal end of tepals; hypogynous glands 4, fleshy, distinct, lobed or fused to form a quadrangular structure (or nectary disk); ovary subsessile, glabrous to densely pubescent, with 2 pendulous, orthotropous ovules, style curved, stigma latero-apical. Fruit a nut or rarely a drupe, indehiscent, 1-2-seeded, subglobose to ovoid; outer mesocarp thin and coriaceous or thick and fleshy; inner mesocarp thin and woody or very thick and sclerous; seeds fleshy, more or less compressed, not winged.

D i s t r i b u t i o n : Endemic to tropical S America, ranging from Venezuela (Bolívar) and central Colombia, to SE Brazil and Bolivia, the center of distribution is in SE Brazil from Minas Gerais to Paraná; 20 species in the Neotropics of which 2 species are known to occur in the Guianas.

E t y m o l o g y : The name *Euplassa* refers to the leaves resembling those of many Leguminosae.

KEY TO THE SPECIES

1 Flower bud apex rounded, lacking a beak (not rostrate); secondary veins wavy . . . . *1. E. glaziovii*
Flower bud apex laterally elongated, forming a short beak (rostrate); secondary veins straight . . . . *2. E. pinnata*

1. **Euplassa glaziovii** (Mez) Steyerm., Fieldiana, Bot. 28: 217. 1951; – *Adenostephanus glaziovii* Mez, Bot. Jahrb. Syst. 12, Beibl. 27: 10. 1890. Type: Brazil, Rio de Janeiro?, Glaziou 13490 (lectotype B, photo 11774 F, isolectotypes C, K).

Euplassa venezuelana Steyerm., Fieldiana, Bot. 28: 217. 1951. Type: Venezuela, Bolívar, Steyermark 60824 (holotype F, isotype NY).

Tree or shrub up to 12 m; young branches orangeish to dull brown, ferrugineous- to rufous- velutinous to short-pubescent, weakly striate; older branches dark brown to grey/ black, glabrous to patchy short yellow/grey-velutinous, or with patches of a white cuticle, striations faint to very prominent, short-fissured; lenticels numerous. Petiole terete, 3.2-10 cm long, 2-3.5 mm in diam. at base, glabrescent to densely whitish or ferrugineous-puberulent, indument closely appressed, rarely spreading; leaf rhachis (3.5-)7-12.2 cm long, terminal appendix 2-7 x 0.5-1 mm; leaflets sessile or subsessile, petiolules 3-10 mm long, 2-4 mm in diam. at base; leaflets 2-4 pairs, opposite or subopposite, drying black, chartaceous to coriaceous, glabrous to sparsely puberulent above, ferrugineous-puberulent in young leaves, shiny or matt above, glabrous to sparsely white or ferrugineous-puberulent beneath, especially on and near midvein at base of leaflet, hairs generally closely appressed, rarely spreading, commonly asymmetrical, narrowly obovate to ovate, less frequently elliptic, often curving; basal pair of leaflets similar to lateral ones but sometimes suborbicular, 3.2-10 x 3.5-5.6 cm; other leaflets 8.4-14.4 x 3.9-5.8 cm; base equal to weakly oblique, decurrent, apex acute to rounded, sometimes truncate in basal leaflets, commonly retuse to emarginate, margin entire, frequently revolute especially at apex or base of blade; venation reticulodromous to cladodromous, with 3-5 pairs of secondary veins, midvein prominent throughout or only on lower half of blade, reaching apex or frequently inconspicuous or forking towards apex, secondaries wavy, faintly prominent, higher order venation conspicuous on both sides of blade. Inflorescence lax to moderately congested, unbranched, axillary or nearly terminal, 2-3.3 cm in diam.; peduncle (2.3-)3.3-5.7 cm long, 1.5-3 mm in diam.; floral rhachis (5.6-) 7.8-24.5 cm long, densely rufous-puberulent to velutino-tomentose; bracteoles heteromorphic, basal ones linear, up to 2 mm long, ferrugineous-hirsute, superior bracteoles smaller, <1 mm long, not recurved; flower-pair peduncle 0.5-3 mm long, rufous-ferrugineous-velutinous to short-pilose; pedicels 1.5-5 mm long. Flower buds not rostrate, 2.25-3 mm broad at apex, ca. 0.75-1 mm broad at midlength, descending to ascending, ferrugineous-velutinous to short-pilose. Flowers 7-11 mm long, all tepals recurving at anthesis; tepals 0.5-0.75 mm across at midlength, not or weakly keeled; anthers subsessile, 1.5-2 mm long; hypogynous glands fused into a quadrangle 0.5-1.25 mm high and 1.25-2 mm across; ovary 2-2.5 mm long, ovoid to pyriform to semi ovoid, rufous-hirsute. Fruiting peduncle ca. 5.8 cm long; infructescence ca. 25 cm long; fruit pedicel 8-9 mm long, ca. 2 mm thick. Young fruit ovoid, 14 x 8.5 mm, smooth.

Distribution: Common on the Gran Sabana of Venezula (Bolívar); generally in shrubland or savanna, and gallery forests; between 1030 and 2000 m alt., usually at lower alt. of approx. 470 m; also collected in Roraima probably on the Guyanan side (GU?:1).

Specimen studied: Guyana or Brazil: Roraima, 1863-64, Appun 1203 (K).

Phenology: Flowering from November to May, collected with young fruits in February.

Note: The type collection of this species, Glaziou 13490, is labeled as from Rio de Janeiro. It is almost certainly an artifact of mislabelling, and represents one of the Schwacke collections pirated by Glaziou.

2. **Euplassa pinnata** (Lam.) I.M. Johnst., Contr. Gray Herb. 73: 42. 1924. – *Roupala pinnata* Lam., Tabl. Encycl. 1: 243. 1792. – *Euplassa meridionalis* Knight, Cult. Prot. 101. 1809, nom. illeg. – *Adenostephanus guyanensis* Meisn. in Mart., Fl. Bras. 5(1): 95. 1855, nom. illeg. Type: French Guiana, Richard s.n. (holotype P, photo 34993 F).   – Fig. 4

22

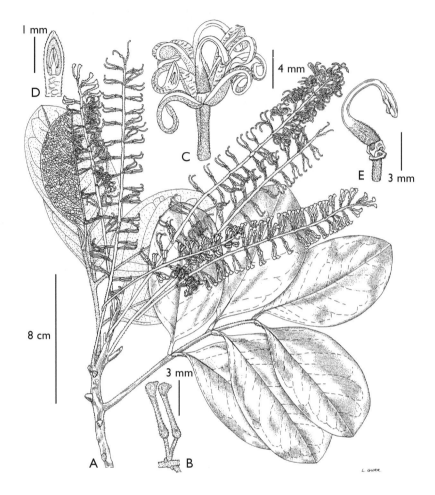

Fig. 4. *Euplassa pinnata* (Lam.) I.M. Johnst.: A, habit; B, flower buds; C, flower; D, anther; E, ovary and nectaries (A-E, Rabelo & Cardoso 2982). Drawing by Linda Gurr.

Tree up to 30 m; young branches orangeish-brown, ferrugineous-velutinous to glabrous, striations narrow, multiple; older branches light brown to black, glabrescent, patchy, sometimes with a thin white coating, striate or small fissured; lenticels numerous to few. Petiole 4.8-12(-20) cm long, terete to laterally compressed, 2-5.5 mm in diam. at base, velutinous or rarely glabrescent with rows of pilose (villous) hairs; leaf rhachis 6-21.7(-45) cm long, terminal appendix 1.5-5 mm long, ca. 1 mm broad, usually absent; leaflets subsessile to short-petiolate, petiolules up to 1.3 cm long, 2.25-2.5 mm in diam. at base; leaflets 3-4(-5) pairs, opposite to almost alternate, drying olive green to black, coriaceous to subcoriaceous, occasionally chartaceous, glabrous above, glabrous to very sparsely white-puberulent beneath, indument if present short and closely appressed and concentrated on venation, frequently asymmetrical, elliptic to broadly elliptic, less commonly obovate or ovate, usually curving; basal pair of leaflets similar to terminal ones, 6.5-14.5(-22.7) x 3.2-5.5(-7.5) cm; other leaflets 10.8-13(-24.5) x 4.3-7.6 cm; base equal to rarely weakly oblique, decurrent, apex acute to obtuse, less frequently rounded, truncate or retuse, especially in basal leaflets, sometimes broadly mucronate, margin entire to denticulate, commonly undulate; venation cladodromous to brochidodromous, with 4-7 pairs of secondary veins, midvein prominent throughout, reaching apex, secondaries straight, faintly prominent, higher order venation always conspicuous beneath, in coriaceous leaves obscure above. Inflorescence moderately lax, occasionally branched, axillary, 2.1-3 cm in diam.; peduncle 2.9-9.4 cm long, 1.5-2 mm in diam.; floral rhachis 12.4–16.5 cm long, ferrugineous to grey, densely appressed-puberulent; bracteoles heteromorphic, basal ones linear, unusually long, up to 4 mm long, light ferrugineous-hirsute, superior ones smaller, broadly triangular, 0.5-1.25 mm long, not recurved; flower-pair peduncle 1-4 mm long, ferrugineous-velutinous; pedicels 1.5-5 mm long. Flower buds 2.5-3 mm broad at apex, rostrate, beak up to 1 mm long, ca. 0.8-1 mm broad at midlength, essentially erect, ferrugineous-villous to sparsely so. Flowers 7-11 mm long, all tepals recurving at anthesis, rarely one remaining erect; tepals 0.5-0.75 mm across at midlength, keeled or not; anthers subsessile, 1-1.2 mm long; hypogynous glands fused into a quadrangle or ring 0.5-0.75 mm high and 1-1.25 mm across; ovary 1.5-2 mm long, narrow, cylindrical-pyriform, square in cross-section, rufous-hirsute to short-pilose, more or less appressed. Infructescence 13.1-14 cm long; fruit pedicel 8-9 mm long, ca. 2 mm thick. Young fruit ovoid, 15-16 x 10-11 mm.

Distribution: Western Amazonian Brazil and French Guiana; in moist forest on terra firme and open forest on sandy soil (SU: 3; FG: 3).

Selected specimens: French Guiana: Cascade Rivière Forêt Noire, BAFOG 1208 (CAY); Saül, Monts La Fumée, Mori *et al.* 15168 (MG, NY).

Vernacular names: French Guiana: lamoussaie blanc, bois grage, bois grage blanc.

Phenology: Flowering September to December, young fruits in December.

2. **PANOPSIS** Salisb. ex Knight, Cult. Prot. 104. 1809.

Type: P. hameliifolia (Rudge) Knight, as 'hameliaefolia' (Roupala hameliifolia Rudge, as 'hameliaefolia')

Shrubs or trees to 25(-40) m tall. Leaves entire, simple, sessile to petiolate, arranged in verticels of 4, offset verticels, decussate, subopposite-decussate or spiral; blades at base sometimes descending petiole on either side, margin not markedly revolute, indument usually present when young, glabrescent, persistent for longer beneath, and above along midvein, less frequently remaining densely pubescent beneath; venation conspicuous to obscure, plane to slightly raised above, more conspicuous and prominent beneath, fractiflexed brochidodromous, frequently eucamptodromous at base, marginal vein weak, derived from ascending lateral veins, midvein raised, plane or more frequently sunken above, prominent and longitudinally ridged beneath, higher order venation reticulate. Inflorescences usually terminal, sometimes lateral or subterminal, unbranched and raceme-like, or branched and panicle-like, bearing lateral inflorescences subtended by leaves which are reduced in size to ligulate and bract-like structures on uppermost inflorescences; lateral inflorescences at times further branched; common bracts subtending flower pairs, caducous, small, ligulate, pubescent on outside, glabrous within; inflorescences subtended by mature leaves or by leaves considerably smaller in size; flowers paired, clustered at short intervals along rhachis (sometimes distinctly whorled) or randomly spread along racemes and branches; pedicels pubescent to pilose. Flower buds elongate or elongate-pyriform, opening apically by elongation of the style. Flowers essentially actinomorphic to slightly zygomorphic, tepals strongly recurved, pubescent to pilose outside, glabrous within; free parts of filaments ribbon-like, adnate to tepals below half-way, or completely free, anthers oblong to elliptic to widely ovate, connective apiculate; hypogynous gland tubular, thin, membranous, 4-lobed, lobes acute to apiculate, eventually shrinking and fragmenting; ovary stipitate, yellow-orange to orange villous outside, glabrous within, style straight-sided to clavate, longitudinally ridged, stigma absent, a specialised stigmatic surface developing at times, ovules 2, collateral, orthotropous, pendulous. Fruit 1-seeded, either globose with rounded to varyingly mucronate apex, surface smooth or scaly, semi-shiny to dull, glabrescent, or fusiform, strongly sutured, corrugate, with indument; pericarpusually with fleshy, granular outer mesocarp, woody, vascularized mid mesocarp, and thin, fleshy inner mesocarp; seed coat thin, often fused to pericarp; seed large, fleshy, rounded, not winged (laterally compressed in *P. rubescens*).

Distribution: Widespread genus of the New World, from Costa Rica and Panama, along the entire Andean chain as far south as Bolivia, and throughout Amazonia, with 1 isolated species in SE Brazil; 2 of the 24 species occur in the Guianas.

Etymology: The name *Panopsis* is derived from the Greek signifying the way in which the petals are "recurved in every way".

KEY TO THE SPECIES

1 Leaves spirally arranged, petioles 1-10 cm long; fruit fusiform-elongated, 2.6-6 x 1.2-3.1 cm, ferrugineous pubescent when young . . . .
. . . . *1. P. rubescens*
Leaves in verticels of 4, sessile or with short petioles 0.1-0.35 cm long; fruit subglobose, 3-5 cm diam., glabrous . . . . *2. P. sessilifolia*

1. **Panopsis rubescens** (Pohl) Rusby, Mem. Torrey Bot. Club 6. 116. 1896. – *Andriapetalum rubescens* Pohl, Pl. Bras. Icon. Descr. 1: 114, t. 91. 1828. Type: Brazil, Goiás, Pohl s.n. (holotype burnt at W, lectotype BR, isolectotype NY, fragment F, possible isotype Pohl Herb. Zuccarini *322*, M). – Fig. 5

Andriapetalum rubescens Pohl var. *acuminatum* Meisn. in A. DC, Prodr. 14: 346. 1856, as 'Andripetalum'. – *Panopsis acuminata* (Meisn.) J.F. Macbr., Publ. Field Mus. Nat. Hist., Bot. Ser. 11: 66. 1931. Type: Brazil, Amazonas, Spruce 2574 (holotype G, photo 7445 F, GH, MO, NY, isotypes BM, C, F, G, GH, K, NY, OXF, not seen at S, B).
Andriapetalum sprucei Meisn. in A. DC., Prodr. 14: 347. 1856, as 'Andripetalum'. Type: Brazil, Amazonas, Spruce 1817 (holotype G, photo 7444 E, F, GE, GH, MO, NY, isotypes BM, GH, K, OXF, P, NY).
Roupala yauaperyensis Barb. Rodr., Vellosia ed. 2. 1: 66, t. 19, fig. A. 1891. Type: [icon] Barb. Rodr., Vellosia ed. 2. 1: t. 19, fig. A. 1891 (designated by Prance *et al.* 2007: 95).
Panopsis cuaensis Steyerm., Ann. Missouri Bot. Gard. 74: 612. 1987. Type: Venezuela, Amazonas, Maguire & Politi 28409 (holotype NY, photo MO-05044 MO).

Tree 6-15 m tall, less frequently treelet or shrub; young branches dense, somewhat appressed, tomentose, villous or puberulous, ferrugineous to pale yellow fading to grey, glabrescent; bark red-brown to grey-brown; lenticels absent or present, small. Leaves spirally arranged, rarely decussate; petioles generally flat, 0.7-10 cm x 1-2 mm at midlength (broadening to 3 mm at

26

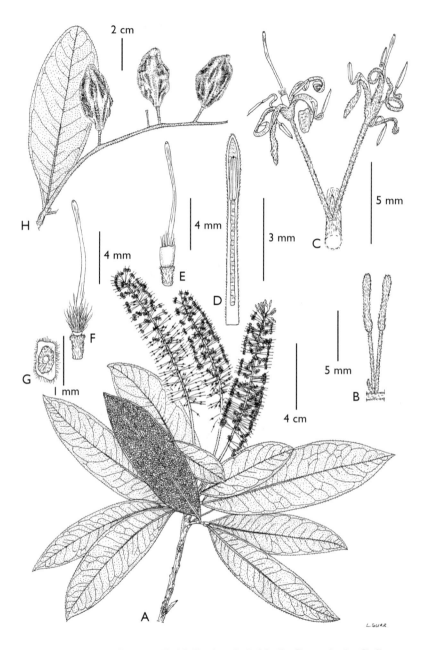

Fig. 5. *Panopsis rubescens* (Pohl) Rusby: A, habit; B, flower buds; C, flowers; D, stamen; E-F, ovary and nectary; G, nectary; H, fruits (A, Spruce s.n.; B-G, Forest Dept. British Guiana 2254; H, Blanco 663). Drawing by Linda Gurr.

base), sericeous, closely appressed, short to long tomentose, villous or puberulous, glabrescent; leaf blade subcoriaceous to coriaceous, closely appressed, long sericeous-pilose above, shorter and less sericeous-pilose beneath, glabrescent both sides except densely persistent hairs on midvein above, ferrugineous to golden or pale yellow fading to grey, elliptic, narrowly elliptic, narrow obovate, oblanceolate, 5-29.5 x 2-9 cm, base symmetrical, acute, at times decurrent, rarely obtuse, apex acute to obtuse, often acuminate, acumen to 15 mm long, margin sometimes slightly revolute; venation not conspicuous, raised above, brochidodromous from lowermost secondary vein, rarely eucamptodromous at base, midvein slightly channelled above, rarely slightly raised, secondary veins 7-16 pairs, remaining straight and parallel almost to leaf margin before ascending steeply, sometimes bifurcating between midvein and marginal vein. Inflorescence unbranched, 5-14 cm x 0.5-2 mm, or branched, 5-25 cm x 1.5-3.5 mm (elongating up to 33 cm after flowering); lateral inflorescences 1-6, spirally arranged, 5-19 cm x 0.5-1.5 mm, yellow-brown to ferrugineous-tomentose, rarely sparsely villous; common bracts (0.3-)1-1.5 mm long, tomentose to pilose; pedicels 4-13 mm long, elongating to 15 mm after flowering, 0.2-0.5 mm broad, short-tomentose to pilose, rarely villous, ferrugineous or yellow-brown. Flowers 3-6.5 mm long; tepals closely appressed, sparsely to densely short sericeous, yellow-brown, rarely ferrugineous; free part of filaments 1.2-3 mm long, adnate 1-2 mm from base of tepals, anthers narrowly oblong to elliptic, 1-1.5 x ca. 0.5 mm; hypogynous gland to 0.5-1.5 mm long, deeply lobed from 1/3 down; ovary hairs reaching 1-2 mm from base, yellow-orange to orange; style 0.2-0.3 mm broad at midlength, clavate, style apex slightly drawn-out, truncate at very tip with small depression centrally. Fruit pedicel 7-13 x 1-2 mm. Fruit resembling a follicle but indehiscent, fusiform but highly variable in shape, 2.6-6.7 x 1.2-3.1 x 0.8-1.7 cm, smooth to irregularly corrugate, ferrugineous rufous or brown, velutinous when young, glabrescent; pericarp (1-)1.5-3 mm thick, outer and mid mesocarp 1-2.5 mm thick, not clearly demarcated, inner mesocarp very thin; seed fleshy, laterally compressed.

Distribution: Colombia, Venezuela, the Guianas, Ecuador, Peru, throughout Amazonian Brazil, and at one locality in Bolivia; growing on river banks, lakesides, islands, in wet savanna, morichales and flooded forest (GU: 10; SU: 5).

Selected specimens: Guyana: Demerara region, Essequibo R., Jenman 979 (K, P, U); Rupununi Distr., Manari, Manari R., Maas *et al.* 3813 (K, MO, NY, U). Suriname: Oelemari, Oelemari R., Wessels Boer 992 (K, NY, U); Sipaliwini Savanna on Brazilian frontier, near rapid in small creek, S headwater of Sipaliwini R., Oldenburger *et al.* 1072 (U). French Guiana: Cayenne, Leblond s.n. (anno 1792) (G).

28

Vernacular name: Guyana: juiba halli

Phenology: Flowering and fruiting all year round, though flowering predominates from May to October and fruiting from October through to April.

2. **Panopsis sessilifolia** (Rich.) Sandwith, Bull. Misc. Inform. Kew 1932: 226. 1932. – *Roupala sessilifolia* Rich., Actes Soc. Hist. Nat. Paris 1: 106. 1792. – *Andriapetalum sessilifolium* (Rich.) Klotzsch, Linnaea 15: 53. 1841, as 'Andripetalum'. Type: French Guiana, Leblond s.n. (224) (holotype P, photo 34994 F, isotype G).

Roupala hameliifolia Rudge, Pl. Guian. Rar. 1: 22, t. 31. 1805, as 'hameliaefolia'. – *Panopsis hameliifolia* (Rudge) Knight, Cult. Prot. 104. 1809, as 'hameliaefolia'. Type: French Guiana, Martin s.n. (lectotype BM, isolectotypes BM, F, K, MO). *Andriapetalum cayennense* Klotzsch ex Meisn. in Mart., Fl. Bras. 5(1): 78. 1855, as 'Andripetalum'. Type: French Guiana, Cayenne, Leprieur s.n. (syntype B photo 11781 F, G, GH, MO, isosyntypes F, K, P, possible isosyntypes E, B); Brazil, Pará, Caripi, Spruce 119 (syntype K) (see Prance *et al.* 2007: 93 for typification problem).

Tree 10-12 m tall; young branches dense short-tomentose, closely appressed pilose, or -villous around axillary bud, glabrescent, yellow, orange or orange-red; bark brown to red-brown; lenticels few to numerous, usually minute. Leaves in verticells of 4, but some paired or spirally arranged especially lower down twigs, sessile; petiole to 5(-7) x 1-4 mm, sparsely pilose or glabrous; leaf blade chartaceous to subcoriaceous, closely appressed-sericeous, pale yellow or grey to yellow-brown, glabrescent, narrow obovate to oblanceolate, occasionally narrowly elliptic or widely obovate, 10-34 x 3-10 cm, base symmetrical, obtuse to acute, less frequently cuneate to ± decurrent, apex acute to rounded, usually acuminate, margin ± revolute; venation conspicuous to obscure, raised both sides, often eucamptodromous at base, brochidodromous further up, midvein plane to sunken above, secondary veins 7-13 pairs, either ascending immediately or straight and parallel before ascending. Inflorescence unbranched or branched, 8.4-31 cm x 1-3 mm, usually with whorls of scale leaves on primary axis (rarely on secondary axes); lateral inflorescences 2-8, at 1-2(-3) nodes, verticillately arranged, 7-30 cm x 0.5-1.5 mm, short orange-brown tomentose to pilose; common bracts 0.5-3 mm long, golden- to orange-brown, tomentose to pilose, or villous; pedicels 4-12(-14) x 0.25-0.4 mm, elongating after flowering, closely appressed sericeous or short tomentose, to pilose, yellow-orange to orange-brown. Flowers 4.5-9 mm long; tepals indument straighter and paler than for pedicels; free part of filaments 1.5-3 mm long, adnate to tepals 1-2(-3) mm from base, anthers oblong, 0.7-1.5 x 0.3-0.7 mm; hypogynous gland 1.5-2.5 mm long, lobed to 0.5-1 mm from top, lobes often acuminate; ovary hairs orange to orange-red villous extending 1-2 mm from base, style 0.2-0.3 mm broad at midlength, ± clavate. Fruit pedicel 6-12 x 2-4 mm. Fruit subglobose, 3-5 cm diam., apex rounded, smooth, glabrous, almost black when immature, lightening to brown or red-brown; pericarp 2-6 mm thick, outer mesocarp 1-5.5 mm thick, thicker in immature fruit, mid mesocarp 0.5-1.5 mm thick, inner mesocarp present when immature, very thin; seed globose, to 2.7 cm diam.; cotyledons thick, fleshy, pale red-brown.

Distribution: Mainly in Venezuela, the Guianas and Brazil (Pará, Amapá and Maranhão), but a few outlying populations in Amazonian Colombia, Ecuador, Peru and Brazil; predominantly at the riparian fringe on terra firme (GU: 17; SU: 14; FG: 16).

Selected specimens: Guyana: Malali, Demerara R., de la Cruz 2738 (F, GH, MO, NY); Kaieteur Plateau, along Potaro R. above Kaiatuk, Maguire & Fanshawe 23372 (A, F, K, MO, NY, RB, U, US). Suriname: Saramacca (no precise locality), Stahel 291 (A, IAN, K, MO, NY, U, UC); Zuid R., 2 km above confluence with Lucie R., Irwin & Prance *et al.* 55878 (F, GH, MO, NY, SP, U). French Guiana: Mana, Melinon 179 (BM, F, P, US); Petit Saut, Sabatier 1034 (CAY).

Vernacular names: Guyana: mahoballi, edauballi. Suriname: foedi iwidale, manari-oedoe. FrenchGuiana: kuyailâ (Wayâpi), meliya (Wayana).

Phenology: Flowering principally from August to November with few specimens collected in December and January, May and July, fruiting January to May and September to November.

Note: Photographs (number 27820) labelled "Types of the Delessert Herbarium" are probably of the true type of *Panopsis sessilifolia*. The specimen is annotated Leblond 224, dated 1792, with an old label determining it as *Roupala sessilifolia*.

3. **ROUPALA** Aubl., Hist. Pl. Guiane 1: 83. 1775.
   Type: R. montana Aubl.

Shrubs, or trees to 25 m; young branches densely hairy, rarely glabrous, glabrescent, lenticellate (occasionally lenticels absent).

Leaves spirally arranged, chartaceous to rigidly coriaceous, long to short petiolate, heterophyllous, compound or pinnatisect when juvenile and sometimes in fertile/ adult state, adult leaves generally simple, glabrous to densely tomentose, quickly glabrescent above, glabrescent or persistent beneath, hairs erect or appressed, whitish or yellowish to ferrugineous or rufous. Compound leaves imparipinnate; leaflets 2-15 pairs, sessile to short-petiolate, asymmetrical, opposite or alternate, margin serrate, rarely entire. Simple leaves short- to long-petiolate, shape various, margin entire to strongly serrate, revolute or not; venation conspicuous to obscure, plane to slightly raised above, generally more prominent beneath, eucamptodromous to semicraspedodromous, occasionally brochidodromous, tertiary venation reticulate, midvein usually reaching leaflet apex. Inflorescences unbranched, pseudo-racemose, axillary or terminal, generally solitary, glabrous to densely tomentose or velutinous; flower-pair axis (peduncle) absent, rarely present; common bracts subtending flower-pairs caducous; flowers pedicellate. Flower buds elongate or elongate-pyriform, opening apically by elongation of style. Flowers actinomorphic; all tepals recurving at right angles at anthesis, glabrous to densely tomentose outside, glabrous within; filaments adnate to perianth to different degrees, generally almost entirely fused, free part ribbon-like, anthers linear to oblong; hypogynous glands 4, fleshy or scale-like, well separated, sometimes fused at base or rarely appearing as a continuous ring; ovary subsessile, glabrous to densely pubescent, sometimes flattened longitudinally and weakly keeled, style erect, claviform, stigma an apical slit, ovules 2, orthotropous, pendulous. Fruit a 1-2-seeded follicle, longitudinally flattened, commonly constricted at base and apex, apex sharp and straight, or strongly curved towards unsutured side, style occasionally persistent; seeds compressed, winged, seed central to wing.

Distribution: Widespread genus of 33 species from Mexico, throughout C America, to Bolivia and Argentina, including the entire Andes, the Guyanan Shield and the Amazon Basin to SE Brazil; from sea-level to 4000 m alt., usually somewhere in the middle of this range; in the Guianas 5 species.

Etymology: The name *Roupala* is derived from the vernacular name "roupale" used in French Guiana.

Note: Especially during the 19th century, the orthographic variant spelling 'Rhopala' was generally used.

KEY TO THE SPECIES

1   Inflorescence, including perianth, velutinous to tomentose, indument obscuring the surface; ovary indument long, bushing outwards; fruit velutinous . . . . *5. R. suaveolens*
Inflorescence glabrous to sparsely pilose or furfuraceous; ovary glabrous or very short-strigose and/or appressed-tomentose; fruit glabrous 2

2   Petioles well defined, 3.5-4 mm broad at midlength; leaves thickly coriaceous, rigid, often splitting when dry, margin ridged beneath, indument very short-strigose, rufous, only distinguishable under high magnification, giving inflorescence and undersurface of young leaves a very dark red tinge . . . . *4. R. sororopana*
Petioles poorly defined, 1-3 mm broad at midlength; leaves subcoriaceous to coriaceous (in some specimens of *R. montana*), margin not ridged beneath, indument absent or pilose to tomentose, without dark tinge 3

3   Fruit strongly curved on sutured side, hemispherical, 1.8-2.2 cm wide, drying dark brown or black when mature; flowers 3.5-5 mm long; pedicels 0.3-0.4 mm broad; ovary glabrous or with sparse indument . . . *2. R. nitida*
Fruit curved more or less equally at sutured and unsutured sides, 0.8-1.6 cm wide, drying beige or grey or black; flowers 5-14 mm long; pedicels 0.3-1 mm broad; ovary hairy (except very rarely in *R. obtusata*) 4

4   Inflorescence sparsely pilose to densely tomentose; fruit not conspicuously ridged nor with a thickened marginal vein; leaves entire or serrate; tertiary venation conspicuous beneath . . . . *1. R. montana*
Inflorescence glabrous; fruit usually veiny due to underlying vascular tissue, mature fruits with a thickened marginal vein; leaves strictly entire; tertiary venation inconspicuous beneath . . . . *3. R. obtusata*

1. **Roupala montana** Aubl., Hist. Pl. Guiane 1: 83. 1775. Type: French Guiana, Serpent Mountain, Aublet s.n. (holotype BM, photo NY).

Roupala pyrifolia Knight, Cult. Prot. 102. 1809. Type: French Guiana, Martin s.n. (holotype BM, isotype MO).
Roupala media R. Br., Trans. Linn. Soc. London 10: 191. 1810. Type: French Guiana, von Rohr s.n. (holotype BM, isotypes P, W, photo F, MO, NY).
Roupala dentata R. Br., Trans. Linn. Soc. London 10: 192. 1810. – *Roupala montana* Aubl. var. *dentata* (R. Br.) Sleumer, Bot. Jahrb. Syst. 76: 173. 1954. Type: Guyana, A. Anderson s.n. (holotype BM).
Roupala complicata Kunth in Humb., Bonpl. & Kunth, Nov. Gen. Sp. ed. qu. 2: 153. 1817. – *Roupala montana* Aubl. var. *complicata* (Kunth) Griseb., Fl. Brit. W. I. 277. 1860. Type: Venezuela, Carabobo, Bárbula, between Valencia [Novam Valenciam] and Portocabello, Humboldt 752 (holotype P, photo 34991 F, GH, MO, isotype B, photo F, GH, MO).

Roupala macropoda Klotzsch & H. Karst. in Klotzsch, Linnaea 20: 473. 1847. Type: Venezuela, Aragua, Karsten s.n. (holotype B, photo F, GH, MO, fragm. NY).

N o t e : A complete synonymy is given in Prance *et al.* 2007: 132-134.

Shrub or tree 1-8(-15) m tall; young branches light brown, short-appressed-pilose, rapidly glabrescent; bark usually light brown when young, dark grey, dark purple to red-grey or red-brown after peeling of outer layer; lenticels numerous, usually inconspicuous. Leaves simple in mature plants, rarely compound, blade chartaceous to very rigid-coriaceous, drying pale grey green, glaucous to mid brown, sparsely pilose, glabrescent, hairs often persistent along lower midvein and below. Simple leaf: petiole 1-6 cm long x 1-2 mm broad at midlength, sometimes canaliculate, sparsely to densely ferrugineous-to grey-velutinous or appressed-tomentose to pilose, glabrescent; blade extremely variable, from narrowly ovate to widely ovate, elliptic to widely elliptic, oblong, and more rarely suborbiculate, 4-14 x 2-9 cm (length:breadth 1.2-3:1), base acute, obtuse, rounded, sharply decurrent, apex acute or obtuse and narrow-attenuate, or rounded, margin revolute or not, entire, undulate, or serrate with 3-22 pairs of teeth; venation obscure to conspicuous, slightly impressed, plane, or slightly raised above, usually conspicuous and prominent below, semicamptodromous, midvein reaching apex, secondary veins 4-9 pairs, marginal vein thick and prominent below, single or double fractiflexed. Compound leaves: 19.7-33 cm long (including 4.5-11.4 cm long petiole); leaflets 4-8 pairs; petiolules lateral leaflets 0-5 mm long, blade lateral leaflets 6.2-12.2 x 1.2-4.3 cm (length: breadth 1.9-6.5:1), base strongly to weakly asymmetrical, cuneate to angustate, acute to rounded, apex acute to attenuate, margin serrate with 3-20 teeth, rarely entire; secondary veins 3-6 pairs; petiolule terminal leaflet 2-2.3(-4) cm long, blade 7.6-11.1 x 2.7-7.6 cm (length:breadth 1.3-3.7:1), base symmetrical to asymmetrical, attenuate, sometimes rounded to almost truncate, apex acute to attenuate, margin serrate with 7-9 pairs of teeth, secondary veins ca. 3 pairs. Inflorescence axillary, occasionally terminal, unbranched, 5-20 cm x 1.7-2.8 mm, indument light brown, brown, short, sparsely; peduncle 0.5-3 cm x 0.8-2.5 mm; sterile bracts few to abundant towards base; common bracts 0.4-1.3 x 0.3-2.5 mm, margin fimbriate, to densely short tomentose outside, glabrous within; flower-pair axis absent; pedicels 1.5-4.5 x 0.3-1 mm. Flower buds 0.8-1.7 mm broad at apex, 0.5-1 mm broad at midlength, square to rounded in section. Flowers 7-9 mm long, glabrous, to very sparsely shortly-appressed-pilose, to densely pilose; filaments 0-0.8 mm long, attached to tepals 2-6.5 mm from base, anthers 1.5-3 x 0.4-0.8 mm; hypogynous glands 0.2-0.6 mm long, fleshy, lobes free; ovary symmetrical or asymmetrical, more curved on one side than on the other (like half a pair), hairs extending to 1-3 mm from base, covering entire ovary or lower on one side, short-sericeous, hairs light brown, orange, ferrugineous to rufous. Infructescence 5-19 cm long, glabrous to puberulent or sparsely tomentose; fruit pedicels 2-5 x 0.8-1.5 mm. Fruit 2-3.5 x 0.8-1.5 cm, both sutured and unsutured sides curved equally or sutured side curved more strongly, base constricted for 3-6 mm, apex not constricted or constricted to 4 mm, including persistent style base, straight or curved until perpendicular to unsutured side, marginal vein inconspicuous to conspicuous, venation sometimes conspicuous on suface, light brown to dark brown, apressed pilose, or densely velutinous when young, hairs light brown to orange-brown to rufous, glabrescent, glabrous at maturity; seeds 1.8-2.5 x 0.5-1 mm.

Distribution: From Mexico, throughout C America, Trinidad and Tobago, and widespread in S America to southern Brazil, Bolivia, Argentina and Paraguay; in moist, tropical, evergreen to dry, deciduous primary and secondary forest and open grassland habitats of savanna, also in gallery forests and forest remnants (GU: 13; SU: 7; FG: 6).

Selected specimens: Guyana: Pakaraima Mts., Mt. Warinatepu, just N of Paruima Mission, Maas *et al.* 5635 (MO, U, UC); Rupununi Savanna, Mt. Shiriri, Jansen-Jacobs *et al.* 524 (AAU, CAY, K, MO, NY, U, US). Suriname: 5 km E of confluence of Lucie and Oost Rs., Irwin *et al.* 55692 (F, K, MO, NY, RB, U); Brazilian frontier, E Sipaliwini Savanna, N slopes of "4-Gebroeders" Mts., Oldenburger *et al.* 145 (NY, U). French Guiana: Bordelaise Savanna, SW of Cayenne, Route du Tour de l'Ile, Cremers 7830 (CAY); no locality, Mélinon s.n. (anno 1845) (P, US).

Vernacular name: French Guiana: roupale.

Phenology: Flowering and fruiting material has predominantly been collected between September and May, with fruiting specimens also relatively abundant for June and July.

Note: This species has 4 varieties (Prance *et al.* 2007: 132), only var. *montana* occurs in the Guianas.

2. **Roupala nitida** Rudge, Pl. Guian. Rar. 1: 26. 1805. Type: French Guiana, Martin 52 (holotype BM, probable isotypes F, MO).

Tree or shrub 1.5-13 m tall, sometimes scandent; branch bark sandy-brown superficially, dark brown beneath, glabrous; lenticels numerous, inconspicuous. Leaves simple, thin-coriaceous, drying pale green to green-brown, matt or shiny above, glabrous; petiole not well defined, 1-3(-5) cm long x 1-2.5 mm broad at midlength, terete towards base, grooved towards blade rugose, glabrous, petiole base often with conspicuous scar at limit with branch (not stipular), branch elevated where petiole emerge; blade narrowly oblong or oblong, rarely elliptic, 10-36.5 x 3.5-10 cm (length:breadth 2-4: 1), base cuneate or acute-decurrent, usually symmetrical, apex predominantly attenuate, less frequently acute, frequently folded, margin entire, revolute towards base; venation ± conspicuous above, slightly raised, midvein narrowly grooved, conspicuous and prominent beneath, semicraspedodromous (at times appearing eucamptodromous), often forming double marginal vein, midvein reaching apex, secondary veins 5-10 pairs. Inflorescence axillary, unbranched, 3.5-18.5 x 1-1.5 cm, sparsely pilose, ferrugineous; peduncle 0.1-1.3 cm x 0.8-1.8 mm; sterile bracts at base abundant and persistent; common bracts 0.8-1 x 0.5-0.6 mm, glabrous within and without except fimbriate margin of ferrugineous to rufous hairs; flower-pair axis absent; pedicels 1.5-3 x 0.3-0.4 mm, sparsely pilose. Flower buds 0.7-0.8 mm broad at apex, 0.4-0.5 mm broad at midlength, sparsely pilose. Flowers 3.5-5 mm long; filaments 0-0.7 mm long, attached to tepals 2.8-4.5 mm from base, anthers 1-1.8 x 0.2-0.3 mm; hypogynous glands 0.2-0.3 mm long, fleshy, commonly appressed or fused together forming a continuous ring; ovary glabrous or very sparse covered with short, appressed, yellow-brown hairs. Infructescence 4.5-11.5 cm long, glabrous or glabrescent; fruit pedicels 3-5 x 1.2-1.5 mm. Fruit 3-3.7 x 1.8-2.2 cm, hemispherical around uppermost (sutured) edge, slightly rounded on lowermost edge, markedly constricted for 4-7 mm at base, apex constricted for 3-4 mm, persistent base of style forming sharp apex, marginal vein prominent, venation conspicuous on surface, pale brown groove 2 mm wide developing at suture, dark brown or black when immature, becoming paler and opening out completely so the fruit sides are in the same plane when dehisced, remaining attached only for short distance along non-sutured side, glabrous; seeds 2-2.3 x 1.6-1.8 cm, wing somewhat coriaceous.

Distribution: French Guiana and Brazil (Pará), along watercourses which drain into the Oyapock and the Amazon rivers, and disjunct in Colombia (Antioquia); riparian trees in seasonally inundated open forest of islands and river shores, on sandy soil, common at campo and forest boundaries (FG: 7).

Selected specimens: French Guiana: bank of Yaroupi R. (affluent of Oyapock R.), upstream from Saut Couéki, de Granville 481 (CAY); bank of Haut Oyapock R., ca. 3 km upstream from Camopi R., Oldeman 3056 (CAY).

Vernacular name: French Guiana: tatukasi (Wayâpi).

Phenology: Flowering from November to February, fruiting from April to June.

Note: Most closely related to *Roupala suaveolens*, *R. nitida* may be seen as the eastern and western extension to the distribution of that species. However, there are many differences, *R. nitida* lacks the dense indumentum which covers most of the parts of *R. suaveolens*, although it may retain sparse, weak hairs on parts of the inflorescence. *Roupala nitida* frequently has a scar at the base of the petiole which does not occur in *R. suaveolens*. The leaf blade of *R. nitida* has a larger length:breadth ratio 2-4:1 vs. 1.1-2.5:1; flowers are considerably smaller in *R. nitida* (3.5-5 mm long, vs. 7-11 mm long). Despite the smaller flowers, fruits of *R. nitida* are larger, in particular in breadth (3-3.7 x 1.8-2.2 cm vs. 1.8-3.2 x 0.9-1.5 cm), and the suture has a wide groove which develops prior to dehiscence which is not seen in *R. suaveolens*.

3. **Roupala obtusata** Klotzsch, Linnaea 15: 54. 1841. Type: Guyana, Ri. Schomburgk 215 (holotype B, photo 11763 F, GH, K, MO).

> *Roupala obtusata* Klotzsch var. *obovata* Huber, Bol. Mus. Goeldi Hist. Nat. Ethnogr. 5: 338. 1909. Type: Brazil, Pará, Lago de Faro, Ducke MG 6913 (lectotype MG) (designated by Prance *et al.* 2007: 159).
> *Roupala angustifolia* Diels, Notizbl. Bot. Gart. Berlin-Dahlem 6: 288. 1915. Type: Brazil, Amazonas, Rio Negro, Manaus, Ule 8840 (holotype B, photo 11746 F, GH, K, MO).

Tree 2-20 m tall, 10-50 cm diam.; branch bark light brown or grey superficially, dark red-grey or grey beneath, glabrous except rufous axillary buds of short, appressed hairs; lenticels numerous, minute. Leaves simple, or rarely with bisected midvein or extralaminal tissue as evidence of lobing, thinly coriaceous, drying pale green, green-brown or brown and usually shiny above, glabrous; petiole not well defined, (0.5-)1-2.5(-3) cm long, 1-2.5 mm broad at midlength, terete towards base, a scar often developing between branch and petiole (not stipular), smooth to slightly rugose, glabrous; blade predominantly narrowly oblong, lanceolate or elliptic, also oblong, oblanceolate, narrowly obovate or widely elliptic, 4.5-21 x 2-7 cm (length:breadth 1.5-5:1), base cuneate, more rarely acute or obtuse decurrent, symmetrical, apex obtuse to rounded, attenuate or acute, margin entire, revolute towards base of blade; venation raised on both surfaces, tertiary venation faint to obscure beneath, appearing eucamptodromous above, semicraspedodromous below, sometimes forming a double marginal vein, midvein reaching apex, secondary veins 5-9 pairs. Inflorescence axillary, rarely lateral, unbranched, 5-12(-17) x 1.5-3.5 cm, glabrous; peduncle 1-3 cm x 1-1.5 mm; sterile bracts few, persistent; common bracts 0.3-0.5(-0.7) x 0.3-0.6 mm, glabrous except for fimbriate margin and apex of orange- to red-brown hairs; flower-pair axis absent, rarely 0.5-1.5 mm long; pedicels 1-2.5(-4.5) x 0.5-0.8(-1) mm, glabrous. Flower buds 1-1.5 mm broad at apex, 0.5-1 mm broad at midlength, glabrous or occasionally with short, weak hairs. Flowers 6-12(-14) mm long; filaments 0-0.5 mm long, attached to tepals 4-9 mm from base, anthers 2-4.5 x 0.3-0.5 mm; hypogynous glands 0.3-0.5 mm long, closely appressed giving the appearance of a ring; ovary somewhat compressed longitudinally, hairs extending (0.7-)1-1.5(-2.5) mm from base of ovary, extending slightly lower on one side, very short, yellowish to light ferrugineous, close appressed, sometimes glabrous. Infructescence (5-)9-15 cm long, glabrous; fruit pedicels 2.5-6 x 1-1.5 mm, glabrous. Fruit 2.8-3.8 x 1.2-1.6 cm, both sides curved more or less equally, less commonly sutured side curved more strongly than unsutured side, base markedly constricted for (3-)5-7 mm, apex straight to curved to an angle perpendicular to unsutured side, constricted only 1-1.5 mm, style persistent forming sharp apex, surface conspicuously ridged, marginal vein generally protruding throughout periphery, beige to black, glabrous even when immature; seeds 2.1-2.4 x 1.2 cm.

Distribution: From 4° N of the equator in Amazonian Venezuela to 15° S in Mato Grosso, Brazil and between 56° and 67° W, and the unknown type locality in Guyana; almost exclusively found in flooded forest on white sand (GU: 1).

Phenology: Flowering and fruiting predominate from April to September as the Amazonian waters reach peak flood, less commonly fertile from October to December, and February to March with no records for December.

4. **Roupala sororopana** Steyerm., Fieldiana, Bot. 28: 220. 1951. Type: Venezuela, Bolívar, Sororopán-tepui, Steyermark 60089 (holotype F, isotypes NY, VEN).

Roupala chimantensis Steyerm., Bol. Soc. Venez. Ci. Nat. 25: 81. 1963. Type: Venezuela, Bolívar, Chimantá Massif, Torono-tepui, on summit above valley of South Caño, Steyermark & Wurdack 1105 (holotype VEN, isotypes MO, NY).
Roupala schulzii Mennega, Proc. Kon. Ned. Akad. Wetensch. C 69: 334. 1966. Type: Suriname, Wilhelmina Mts., Julianatop, Schulz LBB 10307 (holotype U, isotypes BBS, L, NY).

Roupala paruensis Steyerm., Ann. Missouri Bot. Gard. 74: 613. 1987. Type: Venezuela, Amazonas, summit at the S-SE edge of Cerro Parú, descent to the tributary of Caño Asisa, opening to the Río Ventuari, Cowan & Wurdack 31378 (holotype NY, isotypes F, VEN).

Shrub or treelet 0.5-8 m tall; branch bark dark red-brown to dark grey, with short longitudinal fissures, mostly glabrous; lenticels absent, rarely few. Leaves simple, rigid-coriaceous, drying pale green above, green-brown beneath, matt or shiny, young leaves with dense indument on both sides, furfuraceous, dark-brown to dark red-brown beneath, sometimes persistent above, glabrescent, sometimes remaining as grey or black patches; petiole well defined, (0.3-)1.5-6 cm long x 3.5-4 mm broad at midlength, grooved towards blade, rugose, glabrescent; blade wide-elliptic or wide-ovate, rarely ovate, (3.5-)6-14 x (1.5-)4-10 cm (length:breadth 1-2:1), base obtuse to rounded to truncate, rarely acute, usually symmetrical, often folded, apex acute, obtuse or rounded, margin entire, very rarely with 2-3 pairs of large crenations, not revolute, rim markedly thickened; venation conspicuous above, often more so than below, midvein narrowly and shallowly grooved, secondary veins slightly impressed to slightly raised above, raised beneath (tertiary venation obscure beneath), semicraspedodromous, branching very irregular, often angular, midvein generally reaching apex; secondary veins 3-6 pairs. Inflorescence axillary, unbranched, overtopping leaves, 10-25 x 1.8-2.5 cm, becoming thickened and woody, furfuraceous, hairs short, close-appressed, dark red-brown, swollen at base; peduncle 0.8-2 cm x 1.5-2.8 mm; sterile bracts few towards base; common bracts 0.6-1 x 0.4-0.8 mm, margin dark red fimbriate; flower-pair axis absent; pedicels 2-4.5 x 0.5-1 mm. Flower buds 1.2-1.8 mm broad at apex, 0.8-1.3 mm broad at midlength, robust, square in cross section, dark brown or purple-grey, hairs swollen-based, short, sparse especially towards apex, dark red. Flowers 6-12 cm long; filaments 0.5-1 mm long, attached to tepals 4.5-9 mm from base, anthers 2-3 x 0.4-0.7 mm; hypogynous glands 0.4-0.6 mm long, lobes in contact at base; ovary hairs extending 2.5-4 mm from base, covering entire ovary and base of style, short-strigose, dark red to red-brown. Infructescence (5-)12.5-14.5 cm long, glabrous or glabrescent; fruit pedicels 3-4.5 x (1.2-)1.5-2.5 mm. Fruit oval, (1.8-)3-3.8 x (1.1-)1.3-1.9 cm, sutured and unsutured sides curved +/- to same degree, marginal vein strong, raised, running around periphery of fruit, base constricted for 3-5 mm, apex a small point, constricted ca. 1.5 mm or less, style sometimes persistent; seeds 1.7-2.7 x 0.7-1 cm.

Distribution: Venezuela (Gran Sabana), on the summits of tepuis of the Chimantá Massif (Sororopán-, Torono-, Uei-, Chimantá-tepui) and also on the summit of Julianatop, 1200 m alt., part of the Wilhelmina Mts. in Suriname (SU: 1).

Phenology: Flowering from November to February and fruiting in January on the Chimantá Massif, flowering in August in the Wilhelmina Mts.

Note: This species is most closely related to *Roupala suaveolens*, but grows at generally higher altitudes. The leaves of *R. soropoana* are more rigid, and have a very different type of indument. Whereas the indument of *R. suaveolens* is velutinous to tomentose, that of *R. sororopana* is furfuraceous giving the underside of leaves a dirty appearance. Measurements for parts of *R. suaveolens* and *R. sororopana* usually fall within the same ranges, though the leaf length of *R. sororopana* is generally shorter, the common bracts are shorter and the ovary hairs reach considerably further from the base of the ovary (hairs short-strigose and close-appressed vs. hairs long, weak, sericeous, dense and bushing-out).
Roupala schulzii and *R. paruensis* have been included in synonymy with *R. sororopana* because the slight variations which originally separated them are unique to single (type) specimens and not considered valid under the broader concept of *R. sororopana* used here. *R. schulzii* was said to be distinguished from *R. chimantensis* by the number and course of secondary veins, and the longer pedicels and shorter perianth lobes, which lack an indument. The type specimen of *R. schulzii* is not different to the normal variation seen within *R. sororopana,* but from a lower altitude than is common in *R. sororopana.*

5. **Roupala suaveolens** Klotzsch, Linnaea 20: 473. 1847. Type: [Guyana] Brazil, Roraima, Ri. Schomburgk 855 (holotype B, photo 11767 F, isotypes GH, MO).

Roupala schomburgkii Klotzsch, Linnaea 20: 474. 1847. Type: Guyana, Ri. Schomburgk 1045 (holotype B, isotype NY).
Roupala suaveolens Klotzsch var. *minor* Meisn. in Mart., Fl. Bras. 5(1): 80. 1855. Type: [Guyana] Brazil, Roraima, Ro. Schomburgk ser. II, 544 (holotype fragm. F, isotypes BM, CGE, OXF).
Roupala pullei Mennega, Recueil Trav. Bot. Néerl. 30: 179. 1933. Type: Suriname, Lucie R., Hulk 333 (holotype U).
Roupala griotii Steyerm., Phytologia 44: 321. 1979. Type: Venezuela, Amazonas, vic. Río Coro-Coro and Yutaje airport, Steyermark et al. 117920 (holotype VEN, isotype F).

Tree 3-10(-20) m, or shrub 1.5-2.5 m tall; young branches yellow-brown to red-brown, velutinous to long-tomentose, becoming short-tomentose and fading to grey, persistent; bark dark grey to red-grey or brown; lenticels few, inconspicuous, orange to brown. Leaves simple, lobed and compound, blade coriaceous, drying brown, matt or shiny, densely yellow-brown or rufous-velutinous above and beneath when young, gradually glabrescent, hairs persisting on base of midvein above, and more generally beneath, fading to grey. Simple leaf: petiole 1-6 cm long x 1.5-4 mm broad at midlength, terete, somewhat canaliculate towards blade, yellow-brown velutinous to tomentose, fading to grey, rarely becoming completely glabrous; blade predominantly ovate, also wide- to narrow-ovate, or narrow-oblong to oblong, 6-21 x 3-12 cm (length:breadth 1.1-2.5:1), base obtuse- or rounded-decurrent, cordate or truncate, symmetrical, folded, apex predominantly attenuate, sometimes acute, obtuse or acuminate, margin entire, or serrate when at transitional stage between simple and compound leaves with 1-3(-6) pairs of serrations, not revolute; venation raised and ± conspicuous on both sides, eucamptodromous to semicraspedodromous, midvein reaching apex, secondary veins 5-8(-12) pairs. Compound leaves: 13-26 cm long (including 2.5-8.5 cm long petiole), 7-19 cm broad; leaflets 1-4 pairs; petiolules lateral leaflets 0-0.8 mm long; blade lateral leaflet 4-11.5 x 1.5-4.2 cm (length:breadth 2-3(-4.5):1), base strongly asymmetrical, broader side of blade facing terminal leaflet, acute to obtuse, at times shortest side of blade acute, longer side obtuse, apex attenuate, margin entire, secondary veins 5-8 pairs; petiolule terminal leaflet 0-3 cm long, blade terminal leaflet 7.5-14 x 3.2-8 cm (length:breadth 1.5-3.5: 1), base obtuse to acute, apex attenuate, margin entire, secondary veins 6-10 pairs; compound leaves of juvenile plants with considerably longer and narrower leaves, long-attenuate at apex and long-cuneate at base. Inflorescence axillary, unbranched, rarely 2 arising in same leaf axil, 7-28 x 1.8-2.8 cm, densely pale brown, yellow-brown to rufous, velutinous or tomentose; peduncle 1-3.5(-6) cm x 1.2-2 mm; sterile bracts few, persistent; common bracts 1-2 x 0.6-1 mm, densely velutinous to tomentose outside, glabrous within; flower-pair axis to 1mm long; pedicels 1.5-5 mm long. Flower buds 1.4-2 mm broad at apex, 0.7-1.2 mm broad at midlength. Flowers 7-11 mm long; filaments 0.3-1 mm long, attached to tepals 5.5-9 mm from base, anthers 1.7-2.6 x 0.3-0.4 mm; hypogynous glands 0.3-0.6 mm long, lobes in contact at base; ovary hairs extending (1-)1.5-2(-2.5) mm from base of ovary, covering entire ovary, long, weak, straight, ferrugineous to rufous, also with sericeous hairs, dense, bushing outwards. Infructescence 9-21 (-34) cm long, densely tomentose; fruit pedicel (3-)4-6 x 1-1.5 mm, tomentose. Fruit 1.8-3.2 x 0.9-1.5 cm, both sutured and unsutured side curved to similar degrees, or sutured side slightly more curved, base constricted for (1-)3-7 mm, apex not constricted to constricting for 2.5-3.5 mm, sharp with persistence of style, straight or curved so apex perpendicular to unsutured side, yellow-brown velutinous, sometimes slowly glabrescent, if so, light brown-grey beneath, marginal vein protruding around periphery; seeds 1.3-1.4 x 0.6-0.9 cm.

Distribution: Venezuela (Amazonas and Bolívar) where rivers drain northwards into the Orinoco system, and Guyana; occasional or rare, between 100 and 1300 m alt., in gallery forest and savanna especially beside fast-flowing water also in inundated savanna (GU: 6; SU: 1).

Selected specimens: Guyana: Mazaruni-Potaro Distr., Mt. Roraima, Arabupu, Forest Dept. 2814 (K, FHO, U); Rupununi Distr., Kamoa R., Toucan Mt., Jansen-Jacobs *et al.* 1628 (MO, U).

Phenology: Flowering from September to May, fruiting predominantly from December to March.

Note: *Roupala schomburgkii* was originally differentiated from *R. suaveolens* by the shortness of the petiole, differing shapes of leaves and leaf bases. However, these are inconsistant characters which vary independently of each other in different specimens. *Roupala griotii* is clearly synonymous with *R. suaveolens*.
Although the locality on the type specimen label of *Roupala suaveolens* is difficult to interpret from the handwriting, according to Schomburgk's Travels in British Guiana 1840-1844, vol. 2, by W.E. Roth (1923), this species was seen in the vicinity of Mt. Amboina, Pirocaima and Camana, now in Roraima State, Brazil. The type of *R. schomburgkii* was from the locality of Mt. Roraima which was ascended from the Brazilian side, hence the change in type localities from Guyana to Brazil.

100. COMBRETACEAE

by

Clive A. Stace[5]

Trees, shrubs or lianes, sometimes mangroves, not spiny. Leaves opposite (or whorled) or spiral (often 'alternate'), petiolate, simple, entire, without stipules, often with a pair of petiolar glands. Indumentum almost always of unicellular, slender, thick-walled, pointed hairs with a distinctive basal compartment ('combretaceous hairs') alone or with glandular hairs of one of two types: short, capitate stalked glands, and subsessile peltate 'scales'. Inflorescences axillary or terminal, capitate to expanded, of simple or paniculate spikes or less often racemes; bracts simple, usually caducuous. Flowers bisexual or bisexual and male in same inflorescence or sometimes dioecious, 4- to 5-merous, actinomorphic or sometimes weakly zygomorphic, epigynous; hypanthium (receptacle) surrounding ovary (lower hypanthium) and extended beyond into saucer- to tube-shaped upper hypanthium bearing stamens and perianth, with 2 prophylls (bracteoles) fused to lower hypanthium in Laguncularieae; sepals 4-5, borne at tip of upper hypanthium, sometimes vestigial, rarely accrescent; petals 4-5, free, alternating with sepals, usually borne at or near tip of upper hypanthium, often small, sometimes conspicuous, or often 0; stamens usually twice as many as sepals, borne inside upper hypanthium usually at two levels, sometimes as many as sepals, antipetalous [or antisepalous] whorl missing, exserted or included, anthers 4-locular, dorsifixed, usually versatile, sometimes adnate; nectariferous disk, continuous or as separate lobes, often present at base of or at some distance up upper hypanthium; ovary 1-locular, ovules (1-)2-7, apical, pendulous, style 1, with usually punctiform stigma. Fruits indehiscent, with dry or spongy to succulent wall, often with 2-5 papery to leathery wings; seed 1, endosperm absent, cotyledons usually 2, dorsiventral and variously folded or twisted (mostly spirally convolute or irregularly complicate), rarely conduplicate.

Distribution: About 500 species in 14 genera, distributed throughout the tropics and subtropics of the whole world; 5 genera occur in the New World and in the Guianas; of the 85 species in the Americas, 31 occur in the Guianas (plus 1 nearby in Venezuela).

[5] Department of Biology, University of Leicester, Leicester LE1 7RH, U.K.

Classification: Between 12 and 23 genera have been recognised in the past, the present author recognises 14 genera (Stace 2007a). The genus *Strephonema* (West Africa) constitutes subfamily STREPHONEMATOIDEAE, the other 13 genera forming the COMBRETOIDEAE. The latter is represented by 2 tribes, 1 of which has 2 subtribes; all 3 of these occur in the Guianas:

Laguncularieae: *Laguncularia*
Combreteae with 2 subtribes:
Combretinae: *Combretum* (including *Thiloa*)
Terminaliinae: *Terminalia*, *Buchenavia* and *Conocarpus*

Terminology and measurements: Unless otherwise stated, leaf length excludes the petiole, and flower length includes the ovary and calyx lobes but excludes the petals, stamens and style, and upper hypanthium length also includes the calyx lobes; upper hypanthium width is taken at junction of hypanthium and calyx lobes in *Combretum*, but includes the calyx lobes in *Terminalia*. The lower hypanthium is taken to include the ovary, any pedicel-like proximal region present, and the whole of any solid distal extension below the expanded upper hypanthium. All flower measurements are taken from boiled material; dried material is obviously smaller. Fruit length includes pseudostipe and beak where present, taken at midline; lateral extent of wing is described as wing-width. Leaf venation terminology follows Hickey (1973); descriptions refer to leaf lower surface unless indicated otherwise.

Note: A full synonymy for all taxa is given in Stace, Fl. Neotrop. Monogr. (in press).

LITERATURE

Alwan, A.R.A. 1983. The taxonomy of Terminalia (Combretaceae) and related genera. Unpublished PhD-thesis, University of Leicester.
Bornstein, A.J. 1989. Combretaceae. In R.A. Howard, Flora of the Lesser Antilles 5: 451-463.
Brandis, D. 1893. Combretaceae. In H.G.A. Engler & K.A.E. Prantl, Die natürlichen Pflanzenfamilien 3(7): 106-130.
De Candolle, A.P. 1828. Prodromus systematis naturalis regni vegetabilis. Combretaceae 3: 9-24.
Eichler, A.W. 1867. Combretaceae. In C.F.P. von Martius, Flora Brasiliensis 14(2): 77-128.
Exell, A.W. 1931. The genera of Combretaceae. J. Bot. 69: 113-128.
Exell. A.W. 1935. Combretaceae. In A.A. Pulle, Flora of Suriname 3(1): 164-177.

Exell, A.W. & C.A. Stace. 1963. A revision of the genera Buchenavia and Ramatuella. Bull. Brit. Mus. (Nat. Hist.), Bot. 3: 3-46.

Exell, A.W. & C.A. Stace. 1966. Revision of the Combretaceae. Bol. Soc. Brot. ser. 2. 40: 5-25.

Görts-van Rijn, A.R.A. 1986. Combretaceae. In A.L. Stoffers & J.C. Lindeman, Flora of Suriname, additions and corrections 3(2): 354-355.

Hickey, L.J. 1973. Classification of the architecture of dicotyledonous leaves. Amer. J. Bot. 60: 17-33.

Kawasaki, M.L. 2002. Combretaceae. In S.A. Mori et al., Guide to the Vascular Plants of Central French Guiana. Part 2. Mem. New York Bot. Gard. 76(2): 224-227.

Pulle, A. 1906. An enumeration of the vascular plants known from Surinam, together with their distribution and synonymy. Combretaceae pp. 341-343.

Stace, C.A. 1965. The significance of the leaf epidermis in the taxonomy of the Combretaceae. 1. A general review of tribal, generic and specific characters. J. Linn. Soc., Bot. 59: 229-252.

Stace, C.A. 1971. The type specimens of the species of Thiloa, Buchenavia and Ramatuella described by Ducke. Taxon 20: 337-343.

Stace, C.A. 1981. The significance of the leaf epidermis in the taxonomy of the Combretaceae: conclusions. Bot. J. Linn. Soc. 81: 327-339.

Stace, C.A. & A.R.A. Alwan. 1998. Combretaceae. In J.A. Steyermark et al., Flora of the Venezuelan Guayana 4: 329-352.

Stace, C.A. 2007a. Combretaceae. In K. Kubitzki, The families and genera of flowering plants 9: 67-82.

Stace, C.A. 2007b. Combretaceae. In G. Harling & L. Andersson, Flora of Ecuador 81: 1-63.

Stace, C.A., Combretaceae. In Flora Mesoamericana (in press).

Stace, C.A., Combretaceae. In Flora Neotropica (in press).

Tan, F.-X., et al. 2002. Phylogenetic relationships of Combretoideae (Combretaceae) inferred from plastid, nuclear gene and spacer sequences. J. Pl. Res. 115: 475-481.

KEY TO THE GENERA

1   Leaves opposite or clearly predominantly so; petals usually present 2
    Leaves spiral or alternate; petals absent 3

2   Mangrove shrub or tree with stilt-roots and often pneumatophores; flowers with pair of small prophylls (bracteoles) fused to lower hypanthium . . . .
. . . . 4. *Laguncularia*
Woody lianas, shrubs or small trees, without stilt-roots and pneumatophores; flowers without prophylls (bracteoles) fused to lower hypanthium . . . . 2. *Combretum*

3   Flowers in small dense spherical inflorescences; fruits densely packed into
     cone-like heads ....*3. Conocarpus*
     Individual flowers clearly visible; fruits not in cone-like heads 4

4   Calyx lobes clearly demarcated; anthers dorsifixed and versatile ....
     ....*5. Terminalia*
     Calyx lobes scarcely developed; anthers adnate to filament ....
     ....*1. Buchenavia*

1. **BUCHENAVIA** Eichler, Flora 49: 164. 1866, nom. cons.
    Type: B. capitata (Vahl) Eichler (Bucida capitata Vahl) [=
    Buchenavia tetraphylla (Aubl.) R.A. Howard]

Trees up to 55 m, taller ones with buttresses, not spiny; only
'combretaceous hairs' present. Leaves spirally arranged, usually
clustered at branchlet tips, often with domatia in secondary vein-axils
(pocket-shaped, except usually bowl-shaped in *B. fanshawei*); usually
with petiolar glands. Inflorescences axillary lax to congested simple
spikes, elongated or subcapitate, usually clustered at branchlet-ends;
bracts very small and caducous. Flowers bisexual, actinomorphic,
sessile, 5-merous; lower hypanthium extended into a distinct distal
pedicel-like 'neck', upper hypanthium cupuliform, neck and upper
hypanthium deciduous before fruiting; calyx lobes absent or scarcely
developed; petals 0; stamens 10, slightly exserted, anthers adnate to
filaments; disk densely pubescent; style free, usually 1-3 mm long,
included to shortly exserted, glabrous. Fruits 5-ridged or ± terete,
variably succulent pseudodrupe, radially symmetrical or rarely
somewhat flattened, ridges sometimes very pronounced.

Distribution: A genus of 20 species, confined to the New World,
extending from Costa Rica and Cuba southwards to Brazil (Rio Grande
do Sul); 13 species occur in the Guianas.

Note: There remain several taxonomic problems in the genus, in which
there are few obvious diagnostic characters. Good fruiting material is
usually sufficient for naming, but sterile material or collections with old
fruits or with flowers often remain unnamed.

Classification: The genus is divided into 2 sections, both
represented in the Guianas:

Buchenavia. Flowers at full anthesis congested into subcapitate spikes;
   rhachis up to 1 cm and usually much less, not or scarcely elongating
   in fruit. *Buchenavia ochroprumna, B. parvifolia* and *B. tetraphylla.*

Dolichostachys. Flowers at full anthesis in more or less elongated spikes; rhachis usually more than 1.5 cm, if less then considerably elongating in fruit. The other 10 species.

LITERATURE

Alwan, A.R.A. & C.A. Stace. 1985. Resurrection of two Eighteenth Century names in American Combretaceae - Studies on the flora of the Guianas 7. Nordic J. Bot. 5: 447-449.

Aublet, J.B.C.F. 1775. Histoire des plantes de la Guiane Françoise 1: 224-225, t. 88.

Howard, R.A. 1983. The plates of Aublet's Histoire des Plantes de la Guiane Françoise. J. Arnold Arbor. 64: 255-292.

Van Heurck, H.F. & J. Müller Arg. 1871. Combretaceae. In H.F. Van Heurck, Observationes botanicae et descriptiones plantarum novarum herbarii van Heurckiani 2: 209-249.

KEY TO THE SPECIES

1 Flowers at full anthesis congested into subcapitate spikes; rhachis up to 1 cm long, usually much less, not or scarcely elongating in fruit 2
Flowers at full anthesis in more or less elongated spikes; rhachis usually more than 1.5 cm long, if less then considerably elongating in fruit 4

2 Fruits rufous-tomentose . . . . *8. B. ochroprumna*
Fruits glabrous to very sparsely pubescent 3

3 Mature leaves chartaceous, up to 4(-5.5) x 2 cm; ovary at flowering glabrous
. . . . *10. B. parvifolia*
Mature leaves coriaceous, 2-10.5 x 1-5 cm; ovary at flowering rufous-pubescent . . . . *12. B. tetraphylla*

4 Fruits abruptly narrowed at apex to usually strongly curved beak 1-2 cm long . . . . *6. B. megalophylla*
Fruits rounded to acute or very shortly beaked at apex 5

5 Leaves, inflorescences and fruits conspicuously rufous-tomentose when young, tomentum persisting to maturity or almost so; higher order venation and areolation very strongly raised on lower leaf surface . . . . . .
. . . . *11. B. reticulata*
Leaves, inflorescences and fruits glabrous to variably pubescent, but not conspicuously rufous-tomentose even when young; higher order venation and areolation scarcely to strongly raised on lower leaf surface 6

6   Fruits 3-6.5 cm long, glabrous and with rough rufous scurfy surface .....
.... *4. B. guianensis*
Fruits rarely as much as 3 cm long, glabrous to pubescent without scurfy
surface 7

7   At least some leaves more than 15 cm long; venation eucamptodromous to
eucamptodromous-brochidodromous 8
Leaves usually all less than 14 cm long, rarely a few up to 17 cm; venation
brochidodromous to brochidodromous-eucamptodromous 10

8   Fruits glabrous .... *1. B. congesta*
Fruits densely or sometimes sparsely pubescent 9

9   Petiole eglandular; leaves coriaceous, usually distinctly 'shouldered' at
apex, with usually irregularly percurrent tertiary veins; fruits distinctly
pseudostipitate .... *7. B. nitidissima*
Petiole biglandular; leaves chartaceous, usually gradually acuminate at apex,
with regularly percurrent tertiary veins; fruits not or slightly pseudostipitate
.... *5. B. macrophylla*

10 Secondary veins 3-5 pairs; domatia conspicuous usually bowl-shaped ....
.... *2. B. fanshawei*
Secondary veins (4-)5-10(-12) pairs; domatia usually present then pocket-
shaped 11

11 Most leaves acute to apiculate at apex .... *9. B. pallidovirens*
Most leaves rounded to subacute at apex 12

12 Leaves often over 10 cm long, usually densely clustered on swollen stem
tips; secondary veins 5-10 pairs; inflorescence rhachis 2-10 cm long ....
.... *3. B. grandis*
Leaves up to 9 cm long, not densely clustered on stem tips; secondary veins
4-7 pairs; inflorescence rhachis 1-5 cm long .... *13. B. viridiflora*

1. **Buchenavia congesta** Ducke, Trop. Woods 90: 24. 1947. Type:
Brazil, Amazonas, Manaus, near Cachoeira do Mindú, Ducke 2003
(lectotype RB, isolectotypes A+GH, IAN, INPA, MG, R, U, US)
(designated by Stace 1971: 341).

Buchenavia longibracteata Fróes, Bol. Técn. Inst. Agron. N. 20: 53. 1950.
Type: Brazil, Amazonas, Rio Vaupés, Cachoeira das Araras, Fróes 21308
(holotype IAN, not seen, isotypes K, NY, US).

Tree, 5-40 m, with large buttresses. Leaves chartaceous to subcoriaceous,
obovate to oblanceolate, 8-23 x 3-9 cm, apex shortly and abruptly
acuminate or apiculate (often with good 'shoulders'), base narrowly
attenuate-cuneate, glabrous except sometimes puberulous on midvein

above, glabrous except puberulous on midvein and secondary veins to puberulous except densely so on major venation below; domatia often present; venation eucamptodromous or eucamptodromous-brochidodromous, midvein moderate, prominent, secondary veins 6-16 pairs, moderately spaced to close, arising at moderately to widely acute angles, curved to slightly curved, prominent, intersecondary veins absent, tertiary veins regularly and often closely percurrent; higher order veins distinct; areolation imperfect, slightly prominent; petiole (0.6-)1-3 cm long, sparsely to densely puberulous, usually conspicuously biglandular. Inflorescence 2.5-14 cm long, spicate; peduncle 1.8-3.7 cm, densely rufous-puberulous; rhachis 5-11 cm, densely rufous-puberulous. Flowers 3-4 mm long; lower hypanthium 2-2.5 mm long, probably glabrous to densely pubescent (pilose in type of *B. longibracteata*), with narrow neck ca. half its length, upper hypanthium 1-1.5 x 2.5-3 mm, glabrous. Fruit glabrous or nearly so even when young, 1.8-2.5 x 0.7-1.5 cm, oblong to oblong-elliptic or narrowly so in side view, more or less terete, apex rounded to acute or apiculate, base rounded and usually shortly pseudostipitate.

Distribution: Western catchment of Amazon river-basin in Brazil, Venezuela, Colombia and Ecuador; the isolated localities in Ecuador and Acre, Brazil, and in Guyana, are outliers but the specimens are quite typical; tropical rainforests, often riverine, on flood-plains or terra firme, at 100-400 m elev.; a single specimen has been seen from the Guianas (GU: 1).

Specimen examined: Guyana: NW-Distr., Hurudiah, Moruca R., (fr), van Andel *et al.* 1734 (NY, U).

Vernacular name: Guyana: wild genip.

2. **Buchenavia fanshawei** Exell & Maguire in Maguire *et al.*, Bull. Torrey Bot. Club 75: 648. 1948. Type: Guyana, Essequibo R., below Tukeit, Potaro R. gorge, Maguire & Fanshawe 23499 (holotype NY, isotypes A+GH, BM, F, K, MO, U, US).

Tree 5-35 m. Leaves coriaceous, obovate to narrowly obovate or elliptic, 3-15 x 0.5-6 cm, apex rounded to acute or shortly acuminate, base narrowly cuneate or slightly decurrent, glabrous except sometimes sparsely pubescent on midvein above, glabrous to sparsely pubescent except pubescent to very sparsely so on midvein below; domatia usually conspicuous, usually bowl-shaped; venation brochidodromous, midvein moderate, prominent, secondary veins 3-5 pairs, distant, originating at widely acute angles, curved distally, prominent, intersecondary veins common, tertiary veins randomly reticulate to irregularly percurrent;

higher order veins and areolation indistinct; petiole 0.6-2.4 cm long, glabrous to pubescent, usually eglandular. Inflorescence 1.3-7 cm long, shortly spicate; peduncle 0.6-3 cm long, sparsely puberulous to pubescent; rhachis 0.7-4 cm, puberulous to pubescent. Flowers 3-4 mm long; lower hypanthium 1.5-2 mm long, densely pubescent on ovary, gradually narrowing to rather thick sparsely pubescent neck longer than ovary, upper hypanthium 1.5-2 x 2.5-4 mm, glabrous or nearly so. Fruit densely tomentellous at first, becoming subglabrous when very old, 1.7-2.5 x 1-1.5 cm, ovate to elliptic or broadly so in side view, more or less terete, apex rounded to acute or rarely apiculate, base rounded or very shortly pseudostipitate.

Distribution: Restricted to Guyana and northern parts of Amazonian Brazil; damp, usually riverine, primary or secondary forest, usually on alluvial or sandy soils on riverbanks, at 5-250 m elev.; apart from the type, 7 paratypes and about 16 other specimens from the Guianas (GU: 25).

Selected specimens: Guyana: New R., Corantyne R., Anderson 749 (FDG, K); Upper Demerara-Berbice Region, Essequibo R., from Monkey Jump to Persaud timber concession, Henkel & Williams 2126 (F, LTR, US); Rupununi Distr., Kuyuwini Landing, Kuyuwini R., Jansen-Jacobs *et al.* 2878 (fr), 2898 (fl) (LTR, U).

Vernacular names: Guyana: fukadi, sand fukadi.

Phenology: Flowering June; fruiting March to October.

3. **Buchenavia grandis** Ducke, Arch. Jard. Bot. Rio de Janeiro 4: 148. 1925. Type: Brazil, Pará, Trombetas R., Oriximiná, Ducke 16976 (lectotype MG, isolectotypes BM, F, G, P, RB, US) (designated by Stace 1971: 339).

Buchenavia huberi Ducke, Bol. Técn. Inst. Agron N. 4: 24. 1945. Type : Brazil, Amazonas, Manaus, near Cachoeira do Mindú, Ducke 1450 (lectotype RB, isolectotypes A+GH, F, IAN, K, MG, NY, R, US) (designated by Exell & Stace 1963: 35).

Tree (5-)15-45 m. Leaves coriaceous, obovate or elliptic-obovate to narrowly so, 2.5-16 x 1.3-7 cm, apex rounded to subacute or sometimes retuse or apiculate, base cuneate, densely rufous-pubescent when very young but soon subglabrous except sparsely pubescent on midvein, especially below; domatia usually present; venation brochidodromous or eucamptodromous-brochidodromous, midvein moderate to strong, prominent, secondary veins 5-10 pairs, moderately spaced, originating at moderately to widely acute angles, curved, prominent, intersecondary veins absent, tertiary veins weakly to strongly percurrent, usually rather prominent; higher order veins often distinct; areolation usually well developed, small, usually rather prominent; petiole (0.5-)1-2.8 cm long, pubescent to sparsely so, usually eglandular but sometimes inconspicuously biglandular. Inflorescence 3-12 cm long, spicate; peduncle 1.2-2.4 cm, puberulous; rhachis 2.2-10 cm, puberulous. Flowers 2.5-3.5 mm long; lower hypanthium ca. 2 mm long, glabrous to densely pubescent, with neck ca. half of length and glabrous to sparsely pubescent, upper hypanthium ca. 1.5 x 2.5-4 mm, glabrous to very sparsely pubescent. Fruit glabrous to densely tomentellous, 2.4-3.5 x 1.1-1.6(-2.5) cm, elliptic in side view, more or less terete, apex rounded to apiculate, base shortly pseudostipitate.

Distribution: Amazon river-basin of Brazil and eastern Ecuador and Peru, in eastern Brazil, just reaching French Guiana; dry forest or scrub, hill-forest, bush-savanna, sometimes near rivers, characteristically on non-inundated often sandy soils; 1 collection from the Guianas (FG: 1).

Specimen examined: French Guiana: Saül, La Fumée, Sabatier 1134 (CAY, LTR).

Phenology: Flowering August to October; fruiting October to June.

Use: Wood yellowish-brown, frequently used for construction.

4. **Buchenavia guianensis** (Aubl.) Alwan & Stace, Nordic J. Bot. 5: 447. 1985. – *Pamea guianensis* Aubl., Hist. Pl. Guiane 946. 1775. – *Terminalia pamea* DC., Prodr. 3: 13. 1828, nom. illeg. Type: French Guiana, Aublet s.n. (holotype BM).

Tree 3-35(-?45) m, with short or no buttresses. Leaves coriaceous, oblanceolate to narrowly obovate-oblong, (6-)13-40(-?50) x (2-)4-9.5 cm, apex long- or short-acuminate or cuspidate to apiculate, base narrowly cuneate and often decurrent, glabrous to very sparsely pubescent at maturity or sometimes more pubescent on major veins particularly below and on margin; domatia absent; venation eucamptodromous, midvein stout, very prominent, secondary veins 8-20 pairs, moderately spaced to distant, arising at moderately to widely acute angles, curved or slightly so or only near base, prominent, intersecondary veins sometimes present, tertiary veins rather weakly percurrent; higher order venation distinct; areolation imperfect to well developed, usually prominent; petiole (0.5-) 2-7 cm long, glabrous, usually eglandular. Inflorescence 7-16 cm long, spicate; peduncle 2-3 cm, pubescent, becoming glabrous in fruit; rhachis 5-13 cm, pubescent. Flowers 3-5 mm long; lower hypanthium 2-3.5 mm long, subglabrous to densely pubescent, rather gradually narrowed to neck more than half the total length; upper hypanthium 1-1.5 x 2.5-3.3 mm, glabrous. Fruit glabrous, dull and with rough rufous scurfy surface, 3-6.5 x 1.3-4 cm, oblong to oblong-elliptic in side view, subterete or 6-angled with 3 angles stronger than alternating 3, not or scarcely succulent, apex rounded to subacute or shortly apiculate, base rounded to obtuse and without pseudostipe.

Distribution: French Guiana and Brazil (Amapá, Pará, Amazonas, Rondônia); forest on terra firme, at ca. 100 m elev.; apart from the type only 3 specimens have been seen from French Guiana (FG: 4).

Specimens examined: French Guiana: Station des Nouragues, basin of Arataye R., Sabatier & Prévost 2582 (CAY, LTR), Sabatier 2283 (CAY, LTR), Poncy 1874 (CAY, LTR, P, U).

Phenology: Flowering September; fruiting September to July.

Note: *Buchenavia guianensis* is a most distinctive species in fruit, the scurfy non-succulent fruits are unique in the genus. The leaves, which dry a characteristic olive-green colour, also differ in several respects (thickness, colour, venation) from those of the other large-leaved sympatric species, *B. macrophylla*, *B. reticulata* and *B. megalophylla*, which are all species of inundated forest.

5. **Buchenavia macrophylla** Eichler, Flora 49: 166. 1866. Type: Brazil, Amazonas, Rio Vaupés, near Panuré, Spruce 2507 pro parte (lectotype BR, isolectotypes BM, C, CGE, G, K, LE, NY, OXF, P, RB, TCD, US, W) (designated by Exell & Stace 1963: 32).

Tree 3-20(-30) m, without buttresses. Leaves chartaceous to subcoriaceous, oblanceolate or narrowly elliptic to narrowly obovate or narrowly elliptic-oblong, 3-35 x 1.5-14 cm, apex long- to short-acuminate to apiculate or rarely rounded, base narrowly attenuate-cuneate, glabrous except often puberulous on midvein and secondary veins especially below; domatia absent or shallow except for dense hair-tufts; venation eucamptodromous or eucamptodromous-brochidodromous, midvein moderate to strong, prominent, secondary veins 6-14 pairs, moderately spaced to distant, arising at moderately acute angles, curved, prominent, intersecondary veins usually absent; tertiary veins regularly and often closely percurrent, higher order veins distinct; areolation imperfect, slightly prominent; petiole 0.6-3.5 cm long, pubescent to sparsely so, biglandular (often conspicuously so). Inflorescence 3-13 cm long, spicate; peduncle 0.6-3.2 cm, puberulous to densely so; rhachis 2-10 cm, densely puberulous. Flowers 2.5-3.5 mm long;

lower hypanthium 1.5-2 mm long, densely pubescent on ovary-bearing part, with glabrous to sparsely pubescent neck ca. half total length; upper hypanthium 1-1.5 x 2.5-4 mm, glabrous or nearly so. Fruit densely fulvous-pubescent, 1-3.3 x 0.6-1.3 cm, elliptic to oblong or narrowly so in side view, more or less terete, apex rounded to subacute or apiculate, base rounded to very shortly pseudostipitate.

Distribution: From Pacific coast of Colombia, Venezuela (upper Orinoco R.), French Guiana, Ecuador, Peru, Brazil (whole Amazon river-basin) south to Mato Grosso at ca. 13° S; riverside and lakeside forest, often inundated but sometimes on terra firme, swamp-forest, flood-plains, on sandy or clayey soil, at ca. 150 m elev.; 3 specimens from the Guianas (FG: 3).

Specimens examined: French Guiana: Camp no. 1, Ouman fou Langa Soula, Upper Marouini R., de Granville *et al.* 9614 (CAY, LTR, U, US); Tampoc R., Service Forestier (BAFOG) 7897 (MG, U); First waterfalls of Marouini R., S of Maripasoula, Moretti 824 (CAY).

Vernacular names: French Guiana: alimimo, ingui-tabaka.

Phenology: Flowering July to April; fruiting most of year.

Note: See note under *B. megalophylla*.

6. **Buchenavia megalophylla** Van Heurck & Müll. Arg. in Van Heurck, Observ. Bot. 2: 211. 1871. Type: Guyana, Demerara (?), without collector (holotype AWH, not seen).

Tree 2.5-45 m. Leaves chartaceous, obovate to oblanceolate, 14-36 x 5-14 cm, apex rounded and apiculate to shortly and abruptly acuminate or cuspidate, base narrowly cuneate and scarcely decurrent, glabrous except often puberulous on midvein and secondary veins especially below; domatia absent or shallow except for dense hair-tufts; venation as in *B. macrophylla*; petiole 1.2-2.3 cm long, puberulous, biglandular. Inflorescence 9-16 cm long, spicate; peduncle 2-3 cm, densely pubescent; rhachis 7-13 cm, densely pubescent. Flowers 3-4.5 mm long; lower hypanthium 2-3 mm long, densely pubescent on ovary-bearing part, with less pubescent rather thick neck much less than half total length; upper hypanthium 1-2 x 2.5-4 mm, glabrous or nearly so. Fruit densely fulvous-tomentellous, 1.8-4.7 x 1-1.4 cm, elliptic in side view, strongly 5-ridged, apex abruptly narrowed to usually strongly curved beak of 1-2 cm long, base rounded or obtuse and abruptly narrowed to 0.2-0.7 cm long pseudostipe.

Distribution: Endemic to the Guianas; lowland riverine forest on sand or alluvium, at least seasonally inundated, at 5-100 m elev.; 20 specimens have been seen (GU: 16; SU: 1; FG: 3).

Selected specimens: Guyana: Potaro-Siparuni Region, Iwokrama Rain Forest Reserve, Kurupukari village, base-camp 2 km N on Essequibo R., Mori & Heald 24347 (LTR, US); Kapo, along trail to Lethem, near Karanambo, N Rupununi savanna, Görts-van Rijn et al. 294 (LTR, U); Upper Demerara-Berbice Region, Berbice R., Melissa Falls, 2 km downstream from Doreen Bank Cr., Mutchnick & Harmon 1279 (LTR, US). Suriname: Tumuc Humac Mts., along Litani R., Acevedo-Rodriguez et al. 6160 (LTR, US). French Guiana: Lower Ouaqui R., 2 km above village of Bacarel, de Granville B-4792 (CAY, LTR, U).

Vernacular names: French Guiana: alimi hudu, kwatabobi, ingitabaka.

Phenology: Flowering August to November; fruiting February to July.

Notes: The fruits of this species are quite unmistakable, but flowering or sterile material is impossible to distinguish with certainty from some specimens of *B. reticulata* or *B. macrophylla*. The leaves of *B. macrophylla* are much more variable than those of *B. megalophylla*, often being smaller and often with rather gradually tapered apices, but they can mimic the very large abruptly narrowed leaves of the latter. These two species are possibly allopatric. All undoubted (i.e. fruiting) material of *B. megalophylla* comes from Guyana, but *B. macrophylla* is not known from Guyana or Suriname, although it definitely occurs in French Guiana. There is flowering material probably referable to *B. megalophylla* from Suriname and French Guiana. Sterile material from French Guiana must remain undetermined, although most of it closely resembles typical *B. megalophylla*, with broadly 'shouldered' leaves. Van Heurck & Müller Arg. stated that the type was sent as *Pamea guianensis* to Mlle. H. Reichenbach, probably from Demerara. Specimens (s.n.) coll. Parker, 1821-1824, and labelled [wrongly] *Pamea guianensis* Aubl. are almost certainly duplicates (probable isotypes CGE, G, K; photographs F, NY, US).

7. **Buchenavia nitidissima** (Rich.) Alwan & Stace, Nordic J. Bot. 5: 449. 1985. – *Terminalia nitidissima* Rich., Actes Soc. Hist. Nat. Paris 1: 109. 1792. – *Myrobalanus nitidissima* (Rich.) Kuntze, Revis. Gen. Pl. 1: 237. 1891. Type: French Guiana, Leblond s.n. (holotype G).

Tree 11-50(+) m, with rounded buttresses to 2.5 m high. Leaves coriaceous, elliptic to elliptic-oblong or elliptic-obovate, (5-)12-23 x (3-)4.5-8.5 cm, apex rounded to subacute or abruptly apiculate, base narrowly attenuate-cuneate, glabrous at fruiting except sometimes sparsely pubescent on midvein below; domatia absent; venation eucamptodromous-brochidodromous, midvein stout, prominent, secondary veins 5-9 pairs, moderately spaced, arising at moderately acute angles, curved, prominent, intersecondary veins absent, tertiary veins rather irregularly percurrent, higher order veins distinct; areolation imperfect, large, slightly prominent; petiole 1.5-3 cm long, glabrous or sometimes sparsely pubescent, eglandular. Inflorescence at fruiting 10-18 cm long, spicate; peduncle 3-7 cm, densely puberulous; rhachis 7-11 cm, densely puberulous. Flowers unknown. Fruit densely appressed-pubescent, 2.8-4 x 1.5-2.5 cm, oblong-elliptic to oblong-obovate in side view, more or less terete, apex rounded, base rounded and abruptly shortly pseudostipitate.

Distribution: Endemic to French Guiana; primary moist tropical forest, at 20-400 m elev.; rarely collected, known from the type and 11 other specimens (FG: 12).

Selected specimens: French Guiana: Oyapock R., Pointe de Saint Paul, Oldeman 2516 (NY, P, U, US); Saül, Mts. La Fumée, Mori *et al.* 15095, 15343 (CAY, LTR, NY), Sabatier 1151 (CAY), 2299 (CAY, LTR, U); Approuague R., river banks above Cr. Maripa, Oldeman 2473 (CAY, NY, U); basin of Camopi R., 1 km from Mt. Belvédère, Sabatier & Prévost 1679 (CAY, LTR).

Vernacular names: French Guiana: amandier sauvage, dindaya-oudou.

Phenology: Fruiting December to May.

Note: The flowers are unknown.

8. **Buchenavia ochroprumna** Eichler, Flora 49: 165. 1866. Type: Brazil, Pará, near Santarém, Spruce 309 (lectotype M, isolectotype FI) (designated by Exell & Stace 1963: 16).

Buchenavia discolor Diels, Verh. Bot. Vereins Prov. Brandenburg 48: 192. 1907. Type: Brazil, Amazonas, on banks of Rio Negro, near Manaus, Ule 5979 (holotype B, destroyed, isotypes G, L, RB).

Shrub or tree 3-12(-30) m. Leaves subcoriaceous, obovate to oblanceolate, 2-9.5 x 1-4.5 cm, apex rounded to retuse or rarely obtuse, base cuneate to narrowly decurrent-cuneate, appressed-pubescent when very young, becoming glabrous except often sparsely pubescent on midvein when mature; domatia present; venation brochidodromous, midvein moderate, slightly prominent, secondary veins 3-7 pairs, distant, originating at moderately acute angles, slightly curved, slightly prominent, intersecondary veins occasionally present, tertiary veins randomly reticulate; higher order veins not distinct; areolation incomplete; petiole 0.4-1.5 cm long, glabrous to sparsely appressed-pubescent, eglandular but base of leaf usually obscurely biglandular. Inflorescence 0.7-3.3 cm long, more or less capitate, with densely grouped flowers; peduncle 0.9-3 cm, rufous pubescent in flower, becoming subglabrous and much thicker in fruit; rhachis 0.2-0.5 cm. Flowers 3-4.5 mm long; lower hypanthium 2-2.5 mm long, abruptly narrowed to thin neck of 0.8-1.2 mm long, densely rufous-pubescent except sparsely so on neck, upper hypanthium 1.2-2 x 2.5-3.5 mm, subglabrous. Fruit densely tomentose, 18-30 x 7-17 mm, ovate in side view, terete or nearly so, usually irregularly and strongly ridged, apex abruptly narrowed to 0.4-1 cm long usually strongly curved beak, base rounded.

Distribution: Mainly confined to the lower Amazon river-basin in Brazil (Pará, Amazonas), but extending sparsely N to French Guiana and Venezuela (Upper Orinoco R.) and W to Colombia; lowland, forest along rivers and by lakes, usually in inundated areas and frequently partly submerged, also on sandy river and lake beaches, up to 120 m elev.; a single specimen has been seen from the Guianas (FG: 1).

Specimen examined: Fench Guiana: Comté R., 50 m above Saut Bief, chemin minier de Bief, 700 m from river-bank, Bena 1313 (U).

Vernacular name: French Guiana: angelin rouge.

Phenology: Flowering June to August and fruiting January to May in Brazil, the Guianan specimen fruiting in January.

Use: Wood for construction, boards.

Note: The single specimen from French Guiana, Bena 1313, resembles most of those from Venezuela in having smooth fruits with a short beak. It is possible that they represent a new species, but some Venezuelan specimens have slightly ridged fruits and, as it has not been possible to link flowering or sterile material to the smooth-fruited plants, at present it is best to identify the latter with *B. ochroprumnea*.

9. **Buchenavia pallidovirens** Cuatrec., Fieldiana, Bot. 27(1): 107. 1950. Type: Colombia, Valle, Costa del Pacífico, Bahía de Buenaventura, Quebrada de San Joaquín, Cuatrecasas 19939 (holotype F; isotypes BM, COL, P, U, US).

Tree, 2.5-35 m. Leaves coriaceous, oblanceolate or narrowly obovate to narrowly oblong-elliptic, 3-16 x 1-3.5(-4) cm, apex rounded and apiculate to acute or acuminate, base narrowly cuneate, subglabrous except sparsely pubescent on main veins when mature; domatia present in secondary vein-axils; venation brochidodromous, midvein moderate, prominent, secondary veins 5-8 pairs, moderately spaced, originating at moderately acute angles but immediately strongly curved hence often appearing to arise narrowly acutely, strongly curved, fairly prominent, intersecondary veins usually absent; tertiary veins consistently but weakly percurrent; higher order veins distinct; areolation complete, very small. petiole 0.5-1.5 cm, pubescent, becoming subglabrous when old, biglandular. Inflorescence 4.5-10.5 cm, spicate; peduncle 1-3.5 cm, pubescent to sparsely so; rhachis 3.5-7 cm, pubescent. Flowers 2.5-3.5 mm; lower hypanthium 1.5-2 mm, narrowed to a short neck, wholly densely pubescent; upper hypanthium 1-1.5 x 2.5-4 mm, pubescent. Fruit glabrous or finely pubescent, 1-1.7 x 0.6-1.3 cm, elliptic in side view, more or less terete, apex subacute to rounded, base rounded to truncate.

Distribution: Peru, Colombia, Venezuela, Guyana and north-western Brazil; lowland or upland, often very humid forests, sandy river terraces, at 0-1700 m elev.; only a single specimen has been seen from the Guianas (GU: 1).

Specimen examined: Guyana: Essequibo, Mabura Hill, along road through secondary Wallaba forest (seedlings/saplings with cotyledons), Polak 316 (NY, U).

Notes: No flowers or fruits have been seen from the Guianas. The collection Polak 316 consists of fallen leaves and fruits as well as seedlings, all found under the parent tree. It is probably *B. pallidovirens*, but the material does not permit certainty. The species could well occur in the Guianas on distributional grounds.

10. **Buchenavia parvifolia** Ducke, Arch. Jard. Bot. Rio de Janeiro 4: 150. 1925. Type: Brazil, Pará, Tapajos R., near Villa Braga, Ducke 17686 (lectotype MG, not seen, isolectotypes G, K, P, S, U, US) (designated by Stace 2007b: 44).

Briefly deciduous or ? sometimes evergreen tree, 3-31 m, with plank buttresses 30 cm wide. Leaves chartaceous, usually obovate or narrowly so or oblong-obovate, on juvenile shoots rhombic or obtrullate, (0.6-)1-4(-5.5) x (0.3-)0.5-2 cm, apex rounded or rarely retuse to obtuse, base narrowly decurrent-cuneate, appressed-pubescent when very young, becoming glabrous except often sparsely pubescent on midvein when mature; domatia present; venation brochidodromous, midvein moderate to fine, prominent to scarcely so, secondary veins (3-)4-6(-8) pairs, moderately spaced to distant, originating at widely acute angles, curved to slightly so, slightly prominent, intersecondary veins common, often almost as evident as secondaries, tertiary veins randomly reticulate; higher order veins not distinct; areolation incomplete; petiole 0.2-1.2 cm long, pubescent, becoming sparsely so, eglandular. Inflorescence 1-2.4 cm long, more or less capitate, with only ca. 3-4 closely grouped flowers; peduncle 1-2 cm, rufous pubescent in flower, becoming subglabrous and much thicker in fruit; rhachis 0.2-0.4 cm. Flowers 2.5-4.5 mm long; lower hypanthium 1.5-2.5 mm long, rather gradually narrowed to neck ca. half total length, glabrous, upper hypanthium 1-1.6 x 2.3-3 mm, glabrous to sparsely pilose. Fruit glabrous, 1-2.6 x 0.7-1.6 cm, elliptic to obovate in side view, more or less terete, apex rounded or obtuse and often shortly apiculate (apiculus rarely up to 2 mm), base rounded to acute and sometimes very shortly pseudostipitate.

Distribution: Amazon river-basin from eastern Ecuador and Peru to Pará and Amapá in Brazil, and scattered N to Venezuela and the Guianas; moist evergreen primary rain-forest, often by rivers, on inundated or firm ground, up to 500 m elev. in French Guiana; 6 specimens from the Guianas (GU: 4; FG: 2).

Selected specimens: Upper Takutu-Upper Essequibo Region, Upper Essequibo R., 1-4 km upstream from mouth of Kuyuwini R., Henkel et al. 3353 (F, LTR, U, US); Potaro-Siparuni Region, Kaieteur Falls National Park, along trail to Tukeit, Hahn et al. 4718 (LTR, US). French Guiana: Saül, Mts. La Fumée, Mori & Boom 15300 (CAY, LTR, NY).

Vernacular name: Guyana: mefichi.

Phenology: Fruiting throughout most of year; flowering rarely and no flowers seen from the Guianas.

Uses: Timber used for building houses; firewood.

11. **Buchenavia reticulata** Eichler, Flora 49: 166. 1866. Type: Venezuela, Amazonas, on the Casiquiare, Vasiva and Pacimoni rivers, Spruce 3453 pro parte (lectotype BR, isolectotypes A+GH, BM, CGE, F, G, K, LE, NY, OXF, P, TCD, W) (designated by Exell & Stace 1963: 18).

Buchenavia pulcherrima Exell & Stace, Bull. Brit. Mus. (Nat. Hist.), Bot. 3: 33. 1963. Type: Guyana, 172 km along Bartica-Potaro road, Fanshawe 1485 = Forest Dept. 4221 (holotype K, isotypes FDG, NY).

Tree 3-35 m, with large buttresses. Leaves chartaceous to subcoriaceous, narrowly elliptic to oblanceolate or narrowly obovate or oblong-obovate, rarely obovate, 4-30 x 1-10(-15) cm, apex obtuse to acute, abruptly acuminate or apiculate, base narrowly cuneate, densely rufous-pubescent to -tomentose all over when very young, pubescence varyingly wearing off with age but usually only sparsely pubescent above at flowering, still conspicuously and densely pubescent below at fruiting, especially on venous system; domatia absent; venation eucamptodromous to eucamptodromous-brochidodromous, with ends of secondary veins often forming a looped submarginal vein, midvein strong, very prominent, secondary veins 5-15 pairs, moderately spaced, arising at moderately acute angles, curved, strongly so at base, very prominent, intersecondary veins absent, tertiary veins percurrent or sometimes weakly so; higher order veins distinct; areolation well developed to imperfect, prominent, forming conspicuous raised reticulum; petiole 0.6-4.4 cm long, sparsely rufous-pubescent to tomentose, weakly biglandular, glands obscured by pubescence. Inflorescence 5-18 cm long, spicate; peduncle 2-4.2 cm, sparsely rufous-pubescent to rufous-tomentose; rhachis 3-14 cm, rufous-pubescent to -tomentose. Flowers 3.5-5 mm long; lower hypanthium 2-3 mm long, abruptly narrowed into long thin glabrous or sparsely pubescent neck ca. as long as rufous-tomentose ovary-bearing part, upper hypanthium 1.5-2 x 2.5-4.5 mm, glabrous. Fruit densely rufous-tomentellous, 1.9-4.5 x 0.8-3 cm, oblong to elliptic or narrowly so in side view, terete to distinctly flattened and irregularly longitudinally ridged, apex rounded (outside Guyana more commonly apiculate) or shortly beaked with beak up to 6 mm long, base rounded to obtuse or very shortly pseudostipitate.

Distribution: Mainly in the western part of the Rio Negro catchment, extending N to the Upper Orinoco R., NE to Guyana and northern Venezuela, and W to the Upper Amazon R. and Caqueta R. in Colombia, Ecuador and Peru; primary and secondary forest, usually in inundated areas and by rivers and lakes but also on terra firme, sometimes remaining in open cleared places; only the type of *B. pulcherrima* and the cotyledon-bearing seedlings of the paratype are known from the Guianas (GU: 2).

Specimens examined: Guyana: Bartica-Potaro road, Fanshawe 1485 = Forest Dept. 4221 (holotype K, isotypes FDG, NY), Fanshawe 1486 = Forest Dept. 4222 (paratype FDG, K).

Vernacular name: fukadi (this name is used for several species of *Buchenavia* and *Terminalia*).

Phenology: Fanshawe 1485 has unripe fruits collected in November, the seedlings of Fanshawe 1486 are dated the same month.

Notes: The species has in most of its range oblanceolate leaves, and beaked fruits, but especially in Guyana and parts of Venezuela large-leaved specimens and beakless fruited specimens occur. See note under *B. megalophylla*.

12. **Buchenavia tetraphylla** (Aubl.) R.A. Howard, J. Arnold Arbor. 64: 266. 1983. – *Cordia tetraphylla* Aubl., Hist. Pl. Guiane 224. 1775. Type: [icon] Aubl., Hist. Pl. Guiane t. 88. 1775 (lectotype, designated by Howard 1983: 266). – Fig. 6

Bucida capitata Vahl, Eclog. Amer. 1: 50. 1797. – *Buchenavia capitata* (Vahl) Eichler, Flora 49: 165. 1866. – *Terminalia capitata* (Vahl) Sauvalle, Anales Acad. Ci. Méd. Habana 5: 409. 1869. Type: Leeward Islands, Montserrat, Ryan s.n. (holotype BM, isotypes C, LE).
Buchenavia ptariensis Steyerm. in Steyerm. *et al.*, Fieldiana, Bot. 28: 423. 1952. Type: Venezuela, Bolívar, between Ptari-tepuí and Sororopán-tepuí, Steyermark 60271 (holotype F, isotype NY).

Semi-evergreen (briefly deciduous) tree 2-25(-50) m, with buttresses when large. Leaves coriaceous when mature but very thin at flowering, narrowly to broadly obovate, 2-11 x 1-5 cm, apex rounded to retuse (rarely obtuse), base narrowly acute and usually decurrent, appressed-pubescent when very young, becoming glabrous except often sparsely pubescent on midvein when mature; domatia present; venation brochidodromous, midvein moderate, slightly prominent, secondary veins (2-)3-8 pairs, moderately spaced, originating at moderately acute angles, curved to slightly so, slightly prominent, intersecondary veins occasionally present, tertiary veins randomly reticulate; higher order veins often distinct; areolation usually complete; petiole 0.4-1.5(-2) cm long, sparsely appressed-pubescent, eglandular. Inflorescence 0.8-3 cm long, more or less capitate, with rather few to many densely grouped flowers; peduncle 0.6-2.6 cm, rufous pubescent in flower, becoming subglabrous and much thicker in fruit; rhachis 0.2-0.5(-1) cm. Flowers 2.5-5 mm long; lower hypanthium 1.5-3 mm long, abruptly narrowed to

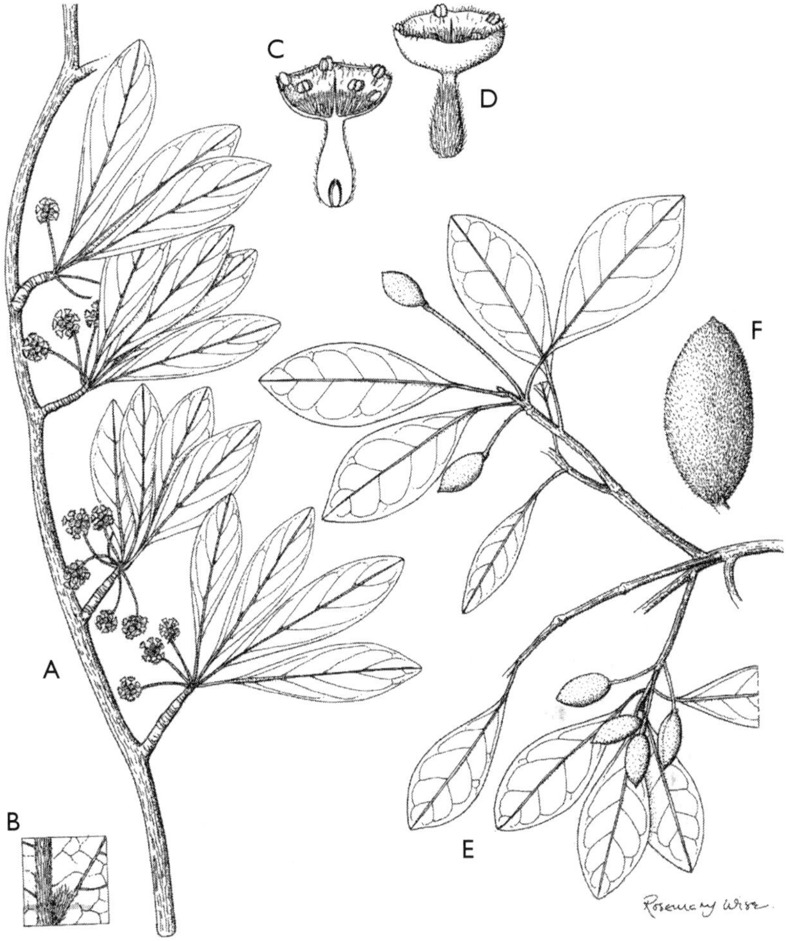

Fig. 6. *Buchenavia tetraphylla* (Aubl.) R.A. Howard: A, flowering shoot; B, domatium in secondary vein axil, lower surface of leaf; C, half flower; D, flower; E, fruiting shoot; F, fruit (A-D, Todzia *et al.* 2331; E-F, Bordenave 816). Drawing by Rosemary Wise.

thin neck of 0.8-1.2 mm long, densely rufous-pubescent except sparsely so on neck, upper hypanthium 1-2 x 2.5-3.5 mm, subglabrous. Fruit glabrous or rarely very sparsely pubescent, (1.2-)1.5-3.5 x (0.7-)1-2.2 cm, ovate to obovate in side view, more or less terete or with 5 low longitudinal ridges, apex acute to rounded and often apiculate, base rounded or rarely shortly pseudostipitate.

Distribution: By far the most widely distributed species of the genus, extending from Cuba and Costa Rica S to Bolivia (ca. 17° S) and Rio de Janeiro, Brazil (ca. 22.4° S); in the Guianas it is usually found in riverside and lakeside forest, inundated or not, up to 500 m elev.; about 27 specimens seen from the Guianas (GU: 2; SU: 7; FG: 18).

Selected specimens: Guyana: Upper Takutu-Upper Essequibo Region, NW Kanuku Mts., 12 km ESE of Nappi, Hoffman 3716 (F, LTR, US). Suriname: Tumuc Humac Mts., Talouakem, Acevedo-Rodriguez *et al.* 5954 (LTR, US); Osembo, Para R., BBS 91 (U); Brokopondo Distr., Forest Reserve Sectie O, tree nr. 30: BW 1312 (U), 3962 (K, U), 4085 (U), 4661 (K, MO, RB, U), 6000 (RB, U), 6261 (NY, RB, U). French Guiana: Montjoly, near Cayenne, Grenand 1734 (CAY, LTR); Roche Koutou, bassin du Haut Marouini R., de Granville *et al.* 9296 (CAY, F, U); Oyapoque R., NE of mouth of Ingarari R., on forested island, Irwin *et al.* 48328 (IAN, K, LTR, MG, NY, U).

Vernacular names: Suriname: gemberhout, gindya-udu, fukadi, fokadi, kanbii, katoelima, katurimja, komanti kwatii, matakki, parakusinja, toekadi, toekoeli.

Phenology: Flowering mostly March to September, fruiting mostly July to March, but both more or less throughout year.

Use: Locally used for timber; "wood is yellow and durable, with close smooth grain".

13. **Buchenavia viridiflora** Ducke, Arq. Inst. Biol. Veg. 2: 63. 1935. Type: Brazil, Amazonas, Manaus, Estrado do Aleixo, Ducke 25023 (lectotype RB, isolectotypes G, K, P, S, U, US) (designated by Exell & Stace 1963: 28).

Shrub or tree 2-55 m, with buttressed trunk up to 1.1 m diam. Leaves chartaceous to coriaceous, obovate to elliptic or narrowly so, (2-)3-9(-10.5) x 1.5-4.5(-5.5) cm, apex rounded to obtusely or subacutely acuminate, base cuneate, subglabrous except sparsely pubescent on midvein, especially below; domatia usually present; venation brochidodromous or eucamptodromous-brochidodromous, midvein moderate, prominent, secondary veins (3-)4-7 pairs, moderately spaced to distant, originating at moderately to widely acute angles, curved to slightly so, prominent, intersecondary veins usually absent, tertiary veins irregularly reticulate to weakly percurrent; higher order veins distinct or not; areolation well developed to inconspicuous; petiole 0.4-2 cm long, pubescent, usually eglandular but sometimes biglandular. Inflorescence

2-7.5 cm long, spicate; peduncle 0.5-2.7 cm, pubescent; rhachis 1.2-5 cm, pubescent. Flowers 3-4.5 mm long; lower hypanthium 1.5-2.2 mm, glabrous to densely pubescent, with glabrous to less pubescent neck ca. half the total length, upper hypanthium 1-1.5 x 1.8-3 mm, sparsely pubescent to subglabrous. Fruit glabrous to densely puberulous, 1.3-2.4 x 0.7-1.3 cm, elliptic in side view, more or less terete, apex rounded to subacute or shortly apiculate,base rounded to obtuse.

Distribution: Amazon river-basin from eastern Colombia, Ecuador and Peru, to French Guiana and NW Pará in Brazil; lowland rainforest on terra firme or inundated by rivers and lakes and in marshes; 3 specimens from the Guianas (FG: 3).

Specimens examined: French Guiana: St. Georges de l'Oyapock, Cr. Cabaret, above and below Saut Emerillon, Grenand 78 (CAY, LTR); Säul, Circuit La Fumée, Mori & de Granville 8794 (CAY, K, LTR, NY); Sinnamary R., Saut l'Autel, Oldeman B-2308 (US).

Phenology: Flowering July to October; fruiting October to July.

Use: Wood yellow, very hard, used for construction.

2. **COMBRETUM** Loefl., Iter Hispan. 308. 1758, nom. cons.
   Type: C. fruticosum (Loefl.) Stuntz (Gaura fruticosa Loefl.)

   Quisqualis L., Sp. Pl.. ed. 2. 556. 1762.
   Type: Q. indica L.
   Cacoucia Aubl., Hist. Pl. Guiane 450, t. 179. 1775.
   Type: C. coccinea Aubl. [= Combretum cacoucia Exell]
   Thiloa Eichler, Flora 49: 149. 1866.
   Type: T. glaucocarpa (Mart.) Eichler (Combretum glaucocarpum Mart.)

Woody lianas to at least 35 m, shrubs, or sometimes small trees to 7 m; 'combretaceous hairs' plus peltate scales or stalked glands present. Leaves opposite or more or less so, sometimes in whorls of three, not clustered at branchlet tips, sometimes with weak pocket-shaped domatia or axillary hair-tufts; without petiolar glands. Inflorescences terminal or axillary, lax to congested, simple to branched spikes, leafless or leafy, spikes often forming compound panicles; bracts usually very small and caducous, sometimes foliaceous. Flowers bisexual, actinomorphic or sometimes weakly zygomorphic, sessile but sometimes lower hypanthium extended into a pseudostipe, 4- or 5-merous; lower hypanthium extended into a short 'neck' or not so, upper hypanthium cupuliform, campanulate or tubiform, deciduous before fruiting; calyx lobes 4 or 5; petals 4 or 5; stamens 8 or 10, usually well exserted, anthers versatile; disk glabrous to densely pubescent; style free, usually exserted, glabrous, or sometimes pubescent proximally. Fruits 4-5-winged or -ridged, actinomorphic, dry or spongy.

Distribution: About 255 species distributed throughout tropical America, Africa and Asia, but not in the Pacific islands, just reaching Australia in the extreme northern tip of Queensland, extending to subtropical regions in S America (to 33° S) and N America (to 30° 35' N); greatest number of species and greatest variation occurs in eastern tropical Africa; in the Americas only 29 species, 9 occur in the Guianas (plus 1 nearby in Venezuela).

Notes: Taxonomic problems in the Guianas concern the distinctions between *C. laxum, C. pyramidatum* and *C. fusiforme.*
Combretum indicum (L.) Jongkind (*Quisqualis indica* L.), from Asia, is a widely planted tropical ornamental liana or shrub, easily distinguished from all Guianan species of Combretum by its scented flowers with a long (38-78 mm) narrow (1.5-3 mm) upper hypanthium and 5 patent pink to red petals 9-20 x 6-13 mm. It is naturalized in some areas of tropical America but apparently not so in the Guianas. In the Flora of Suriname (Exell 1935: 170-171) this species was included but the 2 cited specimens from Paramaribo were cultivated examples, like others since.

Classification: The genus is divided in 3 subgenera and 52 sections, of which 2 and 11 respectively occur in America, and 2 and 5 respectively in the Guianas:
Subgenus 1. **Combretum**. Scales present, stalked glands usually absent; flowers 4-merous, petals glabrous; fruits 4-winged or -ridged. (Sections 1-3).
 Section 1. **Thiloa**. Upper hypanthium deeply cupuliform with scarcely developed calyx lobes; petals absent; stamens 4, scarcely or not exserted; ovules 2; fruits 4-ridged; 'combretaceous hairs' present only inside flowers; scales large and conspicuous. *Combretum paraguariense.*
 Section 2. **Combretum**. Upper hypanthium cupuliform to broadly infundibuliform; petals 4, from shorter than to about as long as calyx lobes; stamens 8, much exceeding calyx lobes; ovules (3-)5-7(-8); fruits 4-winged; 'combretaceous hairs' very sparse to abundant; scales large and conspicuous. *Combretum fruticosum, C. rohrii, C. rotundifolium.*

Section 3. **Combretastrum**. Upper hypanthium cupuliform, without distinct upper and lower regions; petals 4, exceeding calyx lobes; stamens 8, well exceeding calyx lobes; ovules 1-2 (-3); fruits 4-winged or -ridged; 'combretaceous hairs' abundant to extremely sparse; scales small. *Combretum fusiforme, C. laxum, C. pyramidatum.*

Subgenus 2. **Cacoucia**. Scales absent, stalked glands present (usually very inconspicuous); flowers mostly 5-merous, petals pubescent; fruits usually 5-winged. (Sections 4-5).

Section 4. **Spinosae**. Upper hypanthium cupuliform to infundibuliform, without distinct upper and lower regions; petals 4-5, exceeding calyx lobes; stamens 8-10, well exceeding calyx lobes; ovules 4-8; fruits 4-5-winged or -ridged; 'combretaceous hairs' sparse to abundant; peltate scales absent. *Combretum spinosum.*

Section 5. **Cacoucia**. Upper hypanthium deeply curved-cupuliform, without distinct upper and lower regions; petals 5, slightly to much longer than calyx lobes; stamens 10, well exceeding calyx lobes; ovules 3-4; fruits 5-ridged; 'combretaceous hairs' sparse to abundant; peltate scales absent. *Combretum cacoucia.*

LITERATURE

Exell, A.W. 1953. The Combretum species of the New World. J. Linn. Soc., Bot. 55: 103-141.

Stace, C.A. 1968. A revision of the genus Thiloa (Combretaceae). Bull. Torrey Bot. Club 95: 156-165.

Stace, C.A. 1969. The significance of the leaf epidermis in the taxonomy of the Combretaceae. III. The genus Combretum in America. Brittonia 21: 130-143.

Valente, M. da C., N. Marquete Ferreira da Silva & D.J. Guimarães. 1989. Morfologia e anatomia do fruto de Combretum rotundifolium Rich. (Combretaceae). Rodriguésia 41(67): 45-51.

KEY TO THE SPECIES

1 Petals pubescent on at least outer surface and margins; flowers mostly 5-merous; scales absent 2
   Petals if present without hairs; flowers 4-merous; scales present (often inconspicuous) 3

2   Flowers 18-33 mm long, weakly zygomorphic; fruit 5-8 cm long . . . .
. . . . *1. C. cacoucia*
Flowers 3.2-4.4 mm long, actinomorphic; fruit 1-1.4 cm long . . . .
. . . . *9. C. spinosum*

3   Stamens 4; petals absent . . . . *5. C. paraguariense*
Stamens 8 or 10, rarely more; petals present 4

4   Flowers up to 5 mm long, not arranged in 'bottle-brush syndrome'; scales
40-60(-70) μm diam., marginal cells 8(-11) 5
Flowers at least 8 mm long, arranged in 'bottle-brush syndrome'; scales
(50-)70-250 μm diam., marginal cells at least 15 7

5   Leaves coriaceous, with usually only midvein prominent on lower surface
or sometimes secondary veins also slightly so . . . . . . *6. C. pyramidatum*
Leaves chartaceous to subcoriaceous, with all veins from midvein to higher
order venation at least slightly prominent on lower surface 6

6   Fruit 4-6.5 cm long, more than 4 x as long as wide . . . . . . *3. C. fusiforme*
Fruit 1.4-4 cm long, less than 4 x as long as wide . . . . *4. C. laxum*

7   Scales on leaves, flowers and fruit golden-yellow . . . . *2. C. fruticosum*
At least most scales on leaves, flowers and fruit reddish 8

8   Flower buds more or less spherical, apex rounded; upper hypanthium (incl.
calyx) distinctly less than 2 x as long as wide; fruit not much longer than
wide . . . . *7. C. rohrii*
Flower buds elongated, apex acute to obtuse; upper hypanthium (incl. calyx)
ca. 2 x as long as wide; fruit usually about 1.5 x as long as wide . . . .
. . . . *8. C. rotundifolium*

1.   **Combretum cacoucia** Exell, Bull. Misc. Inform. Kew 1931: 469.
1931. – *Cacoucia coccinea* Aubl., Hist. Pl. Guiane 450. 1775, non
Combretum coccineum (Sonn.) Lam. 1785. Type: French Guiana,
bank of Sinamary R., Aublet s.n. (holotype P-JU, isotypes BM, P-
LA, W).   – Fig. 7

Woody liana to 20 m, stem to 10 cm diam., sometimes shrub to 2.5 m or
treelet to 4 m; 'combretaceous hairs' and stalked glands present. Leaves
opposite or more or less so, usually alternate on flowering axis just below
spike, chartaceous to subcoriaceous, elliptic- to oblong-ovate, 3.5-22 x
1.6-10 cm, apex shortly acuminate, base rounded to shallowly cordate,
usually minutely verruculose along all veins, especially on veinlets
below, sparsely to very sparsely pubescent but sometimes moderately so
on veins especially below, stalked glands frequent on veins below and
sparse above; venation usually eucamptodromous or eucamptodromous-

Fig. 7. *Combretum cacoucia* Exell: A, flowering shoot; B, flower bud, with bract; C, flower; D, half-flower, proximal part; E, half-flower opened out, distal part; F, floral diagrams, one with 3 supernumerary stamens; G, petal; H, anthers in two views; I, pollen grains. Reproduced from Eichler, Flora Brasiliensis 14(2), 1867.

brochidodromous, secondary veins 5-9 pairs; petiole 0.3-0.8 cm long, pubescent to densely so, with sparse to frequent stalked glands. Inflorescence simple, terminal, 18-50(-70) cm long; stems leafy to base of spike, rhachis very stout, densely fulvous-pubescent to -tomentose, with sparse to frequent stalked glands; bracts leafy, conspicuous, up to 4.5 x 1.5 cm at lower nodes. Flowers 5-merous, 18-33 mm long, weakly zygomorphic; lower hypanthium 6-14 mm long, pedicel-like region 3-7 mm long, narrowed to neck, densely pubescent, stalked glands very sparse to frequent, upper hypanthium deeply curved-cupuliform to -infundibuliform, 10-19 x 3.5-15 mm, pubescent to densely so outside, very sparse to frequent stalked glands, pubescent to densely so inside; calyx lobes erect, triangular to narrowly so, 2.5-6 mm long, apex acute; petals 5, erect, obovate to narrowly so or elliptic to broadly so, 6-16.4 x 4-5.8 mm, at anthesis slightly to far exserted, apex acute to subacute, pubescent on both surfaces; stamens 10, well exserted, filaments 19.5-29.5 mm long; disk thick, closely applied to style base, densely pubescent at margin, free portion ca. 1.5-2 mm long; style 26.5-44.5 mm long, exserted about as far as filaments, glabrous or pubescent basally; ovules 3-4. Fruit densely pubescent and with frequent stalked glands, becoming subglabrous with age, 5-8 x 1.7-2.8 cm, elliptic in side view, apex gradually tapered to narrowly rounded to obtuse, base gradually and narrowly tapered to pseudostipe of ca. 1-3 cm long, ridges 5, strong but fairly narrow and abruptly edged. Stalked glands 50-110 μm long, stalk uniseriate, 6-14 cells long, obovoid head 25-35 x 20-25 μm.

Distribution: C America from Belize and Guatemala to Panama and just into Colombia, and along the Atlantic side of S America from eastern Venezuela to just south of the Amazon river-delta in north-eastern Pará (Brazil), with a disjunction across most of Colombia and Venezuela for at least 1600 km; primary inundated forest, swamp forest, riverside or lakeside forest, sometimes in or just behind the mangrove zone, also secondary forest by roads and in scrubby savanna, at 0-150 m elev.; about 105 specimens from the Guianas (GU: 40; SU: 27; FG: 38).

Selected specimens: Guyana: Upper Demerara-Berbice Region, Berbice R., 20-30 km SW of Torani Canal, Pipoly et al. 11717 (LTR, U, US); Demerara, Kamuni R., tributary of Demerara R., upstream from Santa Mission, Görts-van Rijn et al., 482 (K, U); Essequibo R., Ro. Schomburgk ser. I, 272 (BM, CGE, G, K, OXF, P, TCD, W). Suriname: Marowijne, along road near Albina, Lanjouw & Lindeman 320 (U); Nickerie, near Apoera, Corentyne R., Heyde 422 (U); Upper Suriname R., Paulus Cr., Mennega 203, 215 (U); Upper Coppename R., Boon 1257 (U). French Guiana: Kaw R., lower and middle courses, de Granville 6852 (CAY, LTR); along road between ferry and Roura, Weitzman & Hahn 247 (LTR, U, US).

Vernacular names: Guyana: jaroomany, yariman, bat-bane, wild almond. Suriname: vreemoesoe-wisi, fremusunoto, sekema, jalimano, yariman, ravetere, puspustere.

Phenology: Flowering most of year; fruiting mostly October to March.

Note: The fruits, which are collected only infrequently, are said to be highly poisonous.

2. **Combretum fruticosum** (Loefl.) Stuntz, U.S.D.A. Bur. Pl. Industr. Invent. Seeds 31: 86. 1914. – *Gaura fruticosa* Loefl., Iter Hispan. 248. 1758. Type: Venezuela, Loefling s.n. (not traced).

Combretum aurantiacum Benth., J. Bot. (Hooker) 2: 222. 1840. Type: Guyana, Essequibo R., Ro. Schomburgk ser. I, 87 pro parte (lectotype BM, isolectotypes G, K, OXF, P, TCD, W) (designated by Stace 2007b: 15); other material of Ro. Schomburgk ser. I, 87 is *C. rotundifolium*.

Woody liana to 30 m (usually much less), stem up to 18 cm diam., in the absence of support a shrub with scandent or procumbent long branches or a shrub 1-5 m; 'combretaceous hairs' (often very scarce) and peltate scales present. Leaves opposite, chartaceous or subcoriaceous, elliptic or sometimes elliptic-oblong to narrowly so, 3.5-19 x 1.5-10 cm, apex usually shortly to long acuminate but sometimes acute or obtuse, base cuneate to rounded or subcordate, often hairless or almost so but sometimes pubescent or even densely so below and sparsely so above, moderately to rather densely yellowish- or golden-lepidote below, sparsely or very sparsely so above; venation usually eucamptodromous-brochidodromous, sometimes eucamptodromous or brochidodromous, secondary veins (5-)6-9(-10) pairs; petiole 0.6-1.5 cm long, glabrous to densely pubescent, contiguously to moderately yellow- to golden-lepidote. Inflorescence unbranched, stout, in opposite pairs in leaf-axils, 4-15 cm long, usually aggregated into terminal racemes up to 25(-40) cm long, at lower nodes in axils of normal leaves (usually fallen by fruiting), at upper nodes usually without or with much reduced subtending leaves, glabrous to rather densely pubescent, densely to contiguously golden-lepidote; flowers borne densely around rhachis but in nature all swept up to vertical position from ± horizontal rhachis, with long exserted stamens forming 'bottle-brush' syndrome. Flowers 4-merous, variable in size and shape, 10.5-17(-20) mm long, varying from glabrous to rufous-pubescent on outside; lower hypanthium 3.5-4.5(-6.5) mm long, 4-angled, pedicel-like region 0.5-1.7 mm long, contiguously golden-lepidote, upper hypanthium 6-12.5(-13.5) x 4-6(-12.5) mm, with narrowly infundibuliform region below disk, 1.5-

4.5(-5) mm long and abruptly demarcated deeply cupuliform region above disk, 3-6(-7.5) mm long excl. calyx lobes, moderately to densely golden-lepidote, often shortly pubescent on inside even when hairless outside; calyx lobes erect, 1.3-2.3(-3) mm long, apex acute to subacute or sometimes obtuse and apiculate, moderately to densely golden-lepidote; petals 4, very rarely absent, elliptic to suborbicular, 1.5-2.5(-3.2) x 0.8-1.5(-2) mm, usually shorter than calyx lobes but sometimes reaching about to their apex, apex obtuse to acute, sometimes apiculate or shortly acuminate, without a basal claw, glabrous, rarely sparsely lepidote outside; stamens 8, far exserted, filaments 12.5-22 mm long; disk densely pubescent at margin with straight or wavy hairs, distinct free margin 0.2-1 mm long; style 13-31.5 mm long, exserted about as far as stamens. Fruit often vinous-red, moderately to densely golden-lepidote especially on body, 1.2-3 x 1-2.9 cm, oblong-elliptic or elliptic to very broadly so, apex obtuse to rounded or retuse and usually distinctly apiculate, base rounded to truncate or slightly retuse with very distinct narrow pseudostipe 0.1-0.55 cm long, wings 4, 0.4-0.9 cm wide. Scales ca. 70-200 µm diam., marginal cells ca. 35-70.

Distribution: Most widespread species of the family in America, occurring from E coast to almost the W coast of S America, and from about 21° N in western Mexico to about 33° S in Argentina and Uruguay, absent in the West Indies apart from Trinidad; a typical forest liana, occupying all sorts of forest: evergreen or deciduous, tall or short, upland or lowland, dry or humid, on slopes or by rivers or lakes or in marshes, virgin, disturbed or secondary, and cut or burned, however, only sparsely scattered in very humid central and lower Amazon river-basin, and notably absent from French Guiana, Suriname and adjacent Guyana, where *C. rohrii* and *C. rotundifolium* are more characteristic; about 10 specimens from the Guianas (GU: 10).

Selected specimens: Guyana: Kanuku Mts., Rupununi R., Bush Mouth near Witaru Falls, Jansen-Jacobs *et al.* 149 (CAY, LTR, U); Mazaruni R., Takutu Cr. to Puruni R., Fanshawe 4799 (K, U).

Phenology: Flowering January to February.

Vernacular names: Guyana: kadiamen. See also under *C. rotundifolium.*

3. **Combretum fusiforme** Gleason, Bull. Torrey Bot. Club 53: 292. 1926. Type: Guyana, Upper Mazaruni R., Kamakusa, de la Cruz 4104 (holotype NY, isotype US).

Woody liana to 30 m; 'combretaceous hairs' and peltate scales present. Leaves opposite, chartaceous, narrowly oblong-ovate, 7-21 x 3-8.3 cm, apex rather gradually shortly to long acuminate, base broadly cuneate to more or less rounded, hairless or almost so, moderately to rather densely lepidote below, sparsely or very sparsely so above, scales usually whitish; venation eucamptodromous-brochidodromous, as in *C. laxum* except secondary veins 10-20 pairs; petiole 0.6-2 cm long, glabrous, inconspicuously lepidote. Inflorescence sparsely to extensively branched, rather stout, in opposite pairs in leaf-axils, 3-15 cm long, usually aggregated into terminal panicles up to 33 cm long, at lower nodes in axils of normal leaves (fallen by fruiting), at upper nodes without subtending leaves, sparsely pubescent, inconspicuously lepidote. Flowers unknown. Fruit rather densely lepidote, 4-6.5 x 0.8-1.3 cm, very narrowly oblong-elliptic to oblanceolate in side view, tapering to narrow apex and base, without distinct pseudostipe, squarish in section with thick short lateral ridges (scarcely wings) running the length of fruit. Scales as in *C. laxum*.

Distribution: Widely disjunct, in Amazonas in extreme S Colombia and NE Peru, and in Guyana and French Guiana; at river margins and in seasonally inundated lowland forest; apart from the type only 3 specimens from the Guianas (GU: 2; FG: 2).

Specimens examined: Guyana: NW-Distr., Hacket Cr., right bank of lower Waini R., Fanshawe 2322 = Forest Dept. 5058 (K, NY, U); French Guiana: no locality, Richard s.n. (P); Cayenne, no other data (G).

Phenology: Fruiting April.

Notes: Exell (1953) placed *C. fusiforme* in synonymy under *C. glabrum* DC., but I see nothing to connect these two taxa or to view *C. glabrum* as other than a synonym of *C. laxum*. The type of *C. fusiforme* lacks inflorescences (only detached fruits), but the Fanshawe specimen has pubescent rhachides, contrasting with the glabrous ones of *C. glabrum*. I have not seen any flowering material that can be linked to *C. fusiforme*, but such material might well be indistinguishable from *C. laxum*. The scattered distribution of *C. fusiforme* might indicate that it is only a very extreme sporadic variant of *C. laxum*.

4. **Combretum laxum** Jacq., Enum. Syst. Pl. 19. 1760. Type: Dominican Republic, Jacquin s.n. (holotype BM).

Combretum obtusifolium Rich., Actes Soc. Hist. Nat. Paris 1: 108. 1792. Type: French Guiana, Leblond s.n. (holotype P).

Combretum puberum Rich., Actes Soc. Hist. Nat. Paris 1:108. 1792. Type: French Guiana, Leblond s.n. (holotype P-LA, isotypes G, P).
Combretum glabrum DC., Prodr. 3: 19. 1828. Type: French Guiana, Cayenne, without collector (holotype G-DC, not seen, two sheets in flower, one annotated by Exell in 1934, IDC microfiche, isotype BM fragment).
Combretum terminalioides Steud., Flora 26: 762. 1843. Type: Suriname, Hostmann & Kappler 1153 (holotype K, isotypes CGE, G, P incl. CN, S, TCD).
Combretum accedens Van Heurck & Müll. Arg. in Van Heurck, Observ. Bot. 2: 234. 1871. Type: Guyana, without collector (holotype AWH, not seen, isotype G).
Combretum brunnescens Gleason, Bull. Torrey Bot. Club 53: 291. 1926. Type: Guyana, NW-Distr., along Amacura R., de la Cruz 3566 (holotype NY, not seen).
Combretum fulgens Gleason, Bull. Torrey Bot. Club 53: 292. 1926. Type: Guyana, Upper Mazaruni R., Kamakusa, de la Cruz 4129 (holotype NY, not seen, isotype US, not seen).
Combretum ulei Exell, J. Linn. Soc., Bot. 55: 132. 1953. Type: Brazil, Amazonas, Rio Juruá, Maray, Ule 5075 (holotype K, isotypes G, L).

Woody liana to 35 m (often much less), stems up to 14 cm diam., in the absence of support a shrub with scandent or decumbent long branches or a shrub to 3 m; 'combretaceous hairs' often very scarce) and peltate scales present. Leaves opposite, chartaceous or subcoriaceous, mostly elliptic to elliptic-oblong, less often narrowly or broadly so, or sometimes ovate-elliptic to narrowly so, (2-)4-24.5 x 1.4-13 cm,apex mostly gradually or abruptly shortly to long acuminate, less often acute or obtuse to rounded, base mostly broadly cuneate to rounded, less often acutely cuneate or subcordate to cordate, mostly hairless or almost so, less often pubescent or even densely so below and sparsely so above, moderately to rather densely lepidote below, sparsely or very sparsely so above, scales usually inconspicuous to naked eye, dark or whitish; venation usually eucamptodromous-brochidodromous, sometimes eucamptodromous or brochidodromous, secondary veins 6-10(-14) pairs, moderately prominent, tertiary veins irregularly to regularly percurrent, slightly prominent, higher order veins distinct, slightly prominent; petiole 0.3-1.1 cm long, glabrous to moderately pubescent, inconspicuously lepidote. Inflorescence sparsely to extensively branched, slender, in opposite pairs in leaf-axils, 3-9.5 cm long, usually aggregated into terminal panicles up to 33 cm long, at lower nodes in axils of normal leaves (often fallen by fruiting), at upper nodes without subtending leaves, glabrous to densely pubescent or rarely tomentose, inconspicuously lepidote; flowers borne densely, less often spaced out, on rhachis. Flowers 4-merous, 2.3-3.8(-5) mm long, glabrous to rather densely pubescent and moderately to densely but usually very inconspicuously lepidote outside; lower hypanthium 0.8-1.8(-2.3) mm long, without or with very short pedicel-like region, more densely pubescent and lepidote than upper hypanthium but sometimes hairless, upper hypanthium deeply cupuliform, 1.4-2(-2.8) x 1.5-3(-3.5) mm, subglabrous to pubescent inside (degree not correlated to pubescence outside); calyx lobes erect, 0.1-0.4 mm long, apex subacute to obtuse; petals 4, patent to reflexed at full anthesis, broadly spathulate with limb transversely oblong-elliptic, 0.8-1.8 x 1-1.5(-2) mm, well exceeding calyx lobes, apex rounded, truncate or irregular, basal claw 0.1-0.5 mm long, glabrous; stamens 8, well exserted, filaments 3-4.7(-6.4) mm long; disk usually sparsely pubescent at margin, without or with narrow free margin up to 0.15 mm long; style (3.1-)4-5.3(-6.6) mm long, usually exserted about as far as stamens. Fruit very variable. Broad-winged fruit: densely but minutely lepidote on body, more sparsely so on wings, glabrous to fairly densely pubescent, especially on body, 1.4-3.8 x 1.3-3.1 cm, broadly elliptic to orbicular in side view, apex retuse, base retuse (often very strongly), slender pseudostipe 0.2-1 cm long, body without spongy tissue, wings 4, thin and flexible, 0.5-1.3 cm wide. Narrow-winged fruit: hairless to tomentose with hairs denser on body, moderately to densely but minutely lepidote especially on body, 1.4-3 (-4) x 0.6-1.2 cm, ovate to narrowly to broadly elliptic or oblong-elliptic in side view, apex acute or acuminate to rounded or truncate, sometimes minutely apiculate, base rounded to narrowly cuneate, thick to narrow pseudostipe 0-0.6 cm long, body much thickened with spongy tissue, wings or ridges 4, stiff, 0.05-0.25 cm wide, sometimes spirally twisted and/or markedly crispate. Scales ca. 40-60 (-70) μm diam., usually whitish, marginal cells 8-11.

Distribution: Most abundant species of the family in America but slightly less widely distributed than *C. fruticosum*, extending from Mexico to ca. 28° S in Argentina; tall evergreen wet or inundated forest, often by rivers or lakes or in swampy ground, but persisting after forest clearance and found in secondary forest, semi-deciduous or semi-evergreen forest, dry forest on slopes and hills, from sea level to 600 m elev.; several hundreds of specimens from the Guianas (many in GU, SU and FG).

Selected specimens: Guyana: Upper Takutu-Upper Essequibo Region, Rupununi area, Surama village, (broad-winged fruits), Acevedo *et al.* 3302 (LTR, U, US); along Soesdyke to Linden Highway, from Timehri Airport to Kuru-Kuru Cr., (narrow-winged corky fruits), Hahn *et al.* 3895 (LTR, U, US); Upper Demerara-Berbice Region, Berbice R., Melissa Falls to the Gate, (intermediate fruits), Mutchnick & Harmon 1295 (LTR, U, US); Lower Waini R., bank of Potaro R., Tumatumari, (flowers), Gleason 338 (BM, K, NY, US), 397 (NY, US); Upper Takutu-Upper Essequibo Region, Kuyuwini R., W of end of trail from Karaudanau and Parabara

Savanna, (fully galled), Clarke 5074 (LTR, U, US). Suriname: Upper
Suriname R., near Goddo, (broad-winged fruits), Stahel 155 (BM, U);
Zuid R., 2 km above confluence with Lucie R., (narrow-winged corky
fruits), Irwin *et al.* 55789 (F, NY, U); Upper Coppename R., Kodji Cr.,
(flowers), Mennega 532 (C, U); below confluence of Oost R. with Lucie
R., (fully galled), Irwin *et al.* 55359 (LTR, NY). French Guiana: future
barrage of Petit Saut, (broad-winged fruits), Billiet & Jadin 4496 (BM,
BR); near Petit Saut on Sinnamary R., (narrow-winged corky fruits),
Prévost 1781 (CAY, LTR); margin of Comté R., Roche Fendée, 60 km
SSW of Cayenne, (intermediate fruits), Mori & Veyret 8948 (LTR, NY);
St. Laurent region, along Route D9 to Mana, (flowers), Skog *et al.* 7456
(CAY, F, LTR, U, US); Cr. Plomb, basin of Sinamary R., (heavily galled),
Bordenave 784 (CAY, LTR).

Vernacular names: Guyana: supple jack, camooraballi. Suriname:
koepirisi.

Phenology: Flowering July to November; fruiting September to May.

Notes: Extremely variable species, particularly in pubescence of leaves
and inflorecences (from densely pubescent to glabrous), leaf shape, and
notably fruit morphology, but these three lines of variation are not
correlated. The fruits vary from having four wide papery wings and a
slim body to having four narrow ridges and a thick spongy body, with all
intermediates. The development of a spongy mesocarp is associated with
water rather than wind dispersal. Galled fruits are very common, where
they occur often all the fruits on a specimen are galled. The commonest
type of gall is a 4-lobed body less than 1 cm long taking the place of each
fruit, and easily mistaken for the latter. In sterile material the very small
scales on the lower leaf surface separate this species from sections
Thiloa and **Combretum**.
See note under *C. fusiforme.*
See also under *C. pyramidatum.*

5. **Combretum paraguariense** (Eichler) Stace, Fl. Ecuador 81: 13.
2007. – *Thiloa paraguariensis* Eichler, Flora 49: 151. 1866. Type:
Brazil, Mato Grosso, near the confluence of the Paraguay and São
Lorenzo rivers, Riedel 734 (holotype LE, isotypes LE, G, K, NY).

Combretum sprucei Eichler in Mart., Fl. Bras. 14(2): 115. 1867. Type:
Brazil, Amazonas, Rio Negro, near S. Carlos, Spruce 3483 (holotype BM,
isotypes BR, G, K, OXF, LE, P, W).

Thiloa inundata Ducke, Trop. Woods 76: 24. 1943. Type: Brazil, Amazonas, Rio Tocantins (trib. of Rio Solimões), Ducke 644 (lectotype RB, isolectotypes F, IAN, MO, NY, R, US) (designated by Stace 1971: 340).

Woody liana to at least 20 m, less often shrub to 5 m; 'combretaceous hairs' (only inside flowers) and peltate scales present. Leaves opposite or rarely in threes, chartaceous to subcoriaceous, elliptic or elliptic-oblong to broadly or narrowly so, 3.5-14.5(-20) x 1.5-7.5(-9) cm, apex abruptly shortly to moderately acuminate, base rounded to broadly cuneate, very sparsely to moderately lepidote below, extremely sparsely to sparsely lepidote above, especially on midvein; venation eucamptodromous to brochidodromous, secondary veins 4-7 pairs; petiole 0.5-2 cm long, sparsely to moderately lepidote. Inflorescence branched or not, in leaf-axils or terminal on lateral branches, 4-15 cmlong, ultimate branches 1-12 cm long and very slender, its axis moderately to densely lepidote. Flowers 3-5.6 mm long; lower hypanthium narrow and pedicel-like above ovary, 1.7-3.6 mm long, without pedicel-like lower region, the only part of plant densely lepidote, upper hypanthium deeply cupuliform, 1.3-2 x 1.6-2.3 mm, moderately lepidote on outside, usually sparsely pubescent inside; calyx lobes erect, broadly triangular, very short, up to 0.6 mm long, apex obtuse to subacute; petals absent; stamens 4, opposite each calyx-lobe, exceeding hypanthium by up to 0.6 mm, filaments ca. 1.5 mm long, each anther with rather inconspicuous caruncle between pollen-sacs on inner side; disk densely pubescent, bulging between bases of filaments, with rather narrow (ca. 0.3-0.4 mm) free margins, leaving very narrow gap around style-base; style 1.7-2.3 mm long, slightly exserted or not so. Fruit sparsely to moderately lepidote, 2.2-5 x 0.7-2 cm, narrowly oblong to narrowly elliptic-oblong in side view, apex obtuse to narrowly retuse and often apiculate, base obtuse to narrowly rounded, pseudostipe 0-0.05 cm long, 4-angled to narrowly 4-winged, stiff ridges 0-0.5 cm wide. Scales ca. 75-150 μm diam., pale brown, marginal cells 25-40.

Distribution: Scattered throughout the Amazon river-basin from its mouth, extending to French Guiana, Venezuela, Colombia, Ecuador, Peru and Bolivia, and S to the Paraguay and São Lorenzo rivers in Mato Grosso (S Brazil); riverine forest, usually seasonally inundated, often along river-banks or by lakes, on sand or clay, usually lowland; 2 collections from the Guianas (FG: 4).

Specimens examined: French Guiana: Oyapock R., Cr. Gabaret, Oldeman B-1531 (CAY, U); Oyapock R., Camopi R., Oldeman 2637 (CAY), 1791, 2652 (CAY, U).

6. **Combretum pyramidatum** Ham., Prodr. Pl. Ind. Occid. 35. 1825. Type: French Guiana, Desvaux s.n. (holotype P).

Combretum laurifolium Mart., Flora 22(1, Beibl.): 64. 1839. Type: Brazil, Amazonas, Barra do Rio Negro, Coari, Martius s.n. (lectotype M) (designated by Stace 2007b: 25).
Combretum phaeocarpum Mart., Flora 24(2, Beibl.): 1. 1841. Type: Brazil, Amazonas, Rio Negro, Martius s.n. (holotype M).
Combretum nitidum Spruce ex Eichler in Mart., Fl. Bras. 14(2): 118. 1867. Type: Brazil, Amazonas, near Barra, Spruce 1482 (lectotype BM, isolectotypes LE, M, OXF, P, TCD, W) (designated by Stace 2007b: 26).

Woody liana to 20 m, in the absence of support a shrub with decumbent long branches or a shrub to 4 m; 'combretaceous hairs' (often very scarce) and peltate scales present. Leaves opposite, coriaceous, narrowly oblong-ovate to broadly elliptic or almost orbicular, 4-20 x 2-12.5 cm, apex gradually or abruptly shortly to long acuminate, acute, obtuse or rounded, base cuneate to subcordate or sometimes cordate, hairless or almost so, moderately to rather densely lepidote below, sparsely or very sparsely so above, scales usually inconspicuous to naked eye, dark or whitish; venation usually eucamptodromous-brochidodromous, sometimes eucamptodromous or brochidodromous, midvein prominent on lower surface, secondary veins 7-17 pairs, close to distant, not to slightly or sometimes moderately prominent, lesser veins not prominent; petiole 0.3-1.4 cm long, glabrous to sparsely pubescent, inconspicuously lepidote. Inflorescence sparsely to extensively branched, slender, in opposite pairs in leaf-axils, 2-7 cm long, usually aggregated into terminal panicles up to 27 cm long, at lower nodes in axils of normal leaves (often fallen by fruiting), at upper nodes without subtending leaves, glabrous to densely pubescent or rarely tomentose, inconspicuously lepidote; flowers borne densely on rhachis. Flowers 4-merous, 2.6-3.5 mm long, moderately to densely but usually very inconspicuously lepidote on outside; lower hypanthium 1.1-1.3 mm long, without pedicel-like region, pubescent to very sparsely so, upper hypanthium deeply cupuliform, 1.5-2.9 x 1.5-3.4 mm, glabrous to sparsely pubescent outside, subglabrous to pubescent inside (degree not correlated to pubescence outside); calyx lobes erect, 0.2-0.3 mm long, apex subacute to obtuse; petals 4, patent at full anthesis, broadly spathulate with limb transversely oblong-elliptic, 0.7-1.5 x 1.1-1.6 mm, well exceeding calyx lobes, apex rounded or irregular, basal claw 0.1-0.3 mm long, glabrous; stamens 8, well exserted, filaments 3.5-4.8 mm long; disk glabrous to densely pubescent at margin, without or with narrow free margin up to 0.15 mm long; style 3.8-5.3 mm long, usually exserted about as far as stamens. Fruit variable, hairless or nearly so, moderately to densely but minutely lepidote especially on body, 1.5-4.5 x 0.7-1.7 cm, ovate or elliptic to narrowly so in side view, apex rounded to acute or acuminate and sometimes minutely apiculate, base rounded to narrowly cuneate, thick to narrow pseudostipe 0-0.6 cm long, body much thickened with spongy tissue, wings (or ridges) 4, stiff, 0-0.3 cm wide, sometimes spirally twisted undulate. Scales as in *C. laxum*.

Distribution: Typical of lowland Amazonas in Brazil, Venezuela and the Guianas, extending W to eastern Ecuador and Peru and S to almost 15° S in Bolivia.; riverside or lakeside primary forest, usually at least seasonally inundated, but sometimes in drier secondary forest or scrub, on clayey or sandy soils; about 42 specimens from the Guianas (GU: 23; SU: 10; FG: 9).

Selected specimens: Guyana: E Berbice-Corentyne Region, 165 km up Canje R., between Ikuruwa and Mora, Gillespie *et al.* 2436 & 2439 (LTR, U, US); Kanuku Mts., Rupununi R., Puwib R., Jansen-Jacobs *et al.* 288 (CAY, K, LTR, U). Suriname: bank of Marowijne R., Lanjouw & Lindeman 2076 (U); bank of Tibiti R., Lanjouw & Lindeman 1861 (K, U). French Guiana: Cr. Plomb, Sinamary R., Loubry 1949 (CAY, LTR); Belligui, near Cr. Beiman, Maroni R., Sabatier 997 (CAY, LTR).

Phenology: Flowering and fruiting mostly as for *C. laxum*.

Note: Distinguished from *C. laxum* by its much thicker coriaceous leaves. These vary in shape and the segregate *C. laurifolium* (apex acuminate) was based on the opposite end of the spectrum from *C. pyramidatum* (apex rounded). *Combretum nitidum*, the original material of which was wholly galled, has leaves intermediate between the two. There is much variation in fruit shape, but in this species only the ridged spongy fruits are present. Galls are common as in *C. laxum*. There must remain some doubt about the specific distinction of *C. laxum* and *C. pyramidatum*, but so far the differences are maintained.

7. **Combretum rohrii** Exell, J. Linn. Soc., Bot. 55: 122. 1953. Type: French Guiana, von Rohr 149 (holotype BM, isotype C).

Woody liana to 30 m, stems to 15 cm diam., or shrub to 2 m; 'combretaceous hairs' (only inside flowers) and peltate scales present. Leaves opposite, chartaceous to subcoriaceous, elliptic, 1.7-10.6 x 0.9-6 cm, apex usually shortly to long acuminate but sometimes acute or obtuse, base cuneate to rounded, without hairs, moderately orange-red-lepidote below, sparsely or very sparsely so above; venation usually eucamptodromous-brochidodromous, sometimes eucamptodromous or brochidodromous, secondary veins 4-8 pairs; petiole 0.4-1 cm long, glabrous, densely to contiguously orange-red-lepidote. Inflorescence unbranched, stout, in opposite pairs in leaf-axils, 5-18 cm long, usually aggregated into terminal racemes up to 56 cm long, at lower nodes in axils of normal leaves (usually fallen by fruiting, often by flowering), at upper nodes usually without or with much reduced subtending leaves, glabrous, densely to contiguously orange-red-lepidote; flowers borne densely around rhachis but in nature all swept up to vertical position from ± horizontal rhachis, with long exserted stamens forming 'bottle-brush' syndrome. Flowers, 4-merous, 8.5-11.5 mm long, without hairs on outside; lower hypanthium 3-3.5 mm long, 4-angled, pedicel-like region 0.5-1 mm long, contiguously orange-red-lepidote, upper hypanthium 5.5-8 x 3.5-5 mm, narrowly infundibuliform region below disk, 2-3 mm long and abruptly demarcated deeply cupuliform region above disk, 2.5-3.5 mm long, excl. calyx lobes, moderately to densely orange-red-lepidote; calyx lobes erect, 1-1.6 mm long, apex acute, moderately to densely orange-red-lepidote; petals 4, elliptic to suborbicular, 1.1-1.5 x 0.7-1 mm, usually falling short of calyx lobes but sometimes reaching ca. to their apex, apex obtuse to acute, sometimes apiculate, base without a claw, glabrous; stamens 8, far exserted, filaments 11.5-14.5 mm long; disk densely pubescent at margin with straight hairs, distinct free margin 0.5-1 mm long; style 14-19.5 mm long, exserted ca. as far as stamens, glabrous; ovules 3-5. Fruit moderately to densely orange-red-lepidote especially on body, 1.2-2.8 x 1.1-2.2 cm, oblong-elliptic or elliptic to broadly so, apex rounded or retuse and sometimes apiculate, base rounded to truncate or slightly retuse, very distinct narrow pseudostipe 0.2-0.5 cm long, wings 4, 0.4-0.7 cm wide. Scales ca. 70-175 μm diam., marginal cells ca. 20-45.

Distribution: Restricted to French Guiana, NE Brazil (eastern Pará and Amapá) and mouth of Amazon R., from ca. 5° N to ca. 4° S; wet or damp forest on terra firme, both in the uplands and by lowland rivers, up to 230 m elev.; 3 specimens seen from the Guianas in addition to the type and 3 old paratypes (see Exell 1953) (FG: 7).

Specimens examined: French Guiana: Cayenne, Martin s.n. (G), Oldeman B-673 (CAY); Saül, Circuit Grand Boeuf Mort, Mori et al., 18625 (NY, U).

Vernacular names: See under C. rotundifolium.

Phenology: Flowering August to January.

8. **Combretum rotundifolium** Rich., Actes Soc. Hist. Nat. Paris 1:
    108. 1792. Type: French Guiana, Leblond 117 (holotype P-LA,
    isotype G).

Combretum aubletii DC., Prodr. 3: 19. 1828. – *Combretum laxum* Aubl.,
Hist. Pl. Guiane 351. 1775, non *C. laxum* Jacq. 1760. Type: [icon] Aubl.,
Hist. Pl. Guiane t. 137. 1775 (here designated).
Combretum magnificum Mart., Flora 24(2, Beibl.): 4. 1841. Type: Brazil,
Pará, Goajurà, Martius s.n. (lectotype M, isolectotype L) (designated by
Stace 2007b: 18).
Combretum guianense Miq., Ann. Mag. Nat. Hist. 11: 12. 1843. Type:
Suriname, Saramacca, Bergendaal, Focke 407 (holotype U, isotype L).
Combretum punctatum Steud., Flora 26: 761. 1843, non *C. punctatum*
Blume 1826. Type: Suriname, Paramaribo, Para district, Hostmann &
Kappler 825 (holotype U, isotypes BM, C, CGE, G, K, LE, OXF, P incl.
CN, TCD, W).

Woody liana to 15 m, shrub to 7 m; 'combretaceous hairs' (often very
sparse) and peltate scales present. Leaves opposite, chartaceous to
subcoriaceous, ovate to elliptic or sometimes broadly or narrowly so, 4-
17 x 2-11 cm, apex acute to acuminate, base cuneate to rounded or
subcordate, usually hairless but sometimes sparsely pubescent on both
surfaces, moderately to densely reddish- to golden-lepidote below, more
or less scaleless to sparsely lepidote above; venation usually
eucamptodromous-brochidodromous, sometimes eucamptodromous or
brochidodromous, secondary veins (5-)6-9(-10) pairs; petiole 0.3-1 cm
long, glabrous, contiguously reddish- to golden-lepidote. Inflorescence
unbranched, stout, terminal or singly or in opposite pairs in leaf-axils, 3-
18(-22) cm long, sometimes aggregated into terminal racemes up to 29
cm long, glabrous or rarely sparsely pubescent, densely to contiguously
reddish-lepidote; flowers borne densely around rhachis but in nature all
swept up to vertical position from ± horizontal rhachis, with long
exserted stamens forming 'bottle-brush' syndrome. Flowers 4-merous,
very variable in size and shape, 11-20.3 mm long, hairless on outside;
lower hypanthium 3.3-4.3 mm long, 4-angled, pedicel-like region 0.6-
1.5 mm long, contiguously reddish-lepidote, upper hypanthium
variously shaped, infundibuliform to bucciniform, 6.3-16 x 3.5-7 mm,
often swollen in region below disk and narrower above disk but flared
near apex (below calyx lobes), 5.1-12 mm excluding calyx lobes,
moderately to densely reddish-lepidote, sparsely to densely pubescent
inside; calyx lobes erect, 1.2-4.3 mm long, apex acute to subacute,
moderately to densely reddish-lepidote; petals 4, elliptic, 1.7-3.1 x 0.8-
1.2 mm, usually falling well short of calyx lobes but sometimes reaching
about to their apex, apex subacute to acute, base without claw, glabrous

but sometimes sparsely lepidote outside; stamens 8, far exserted, filaments 18.5-29.5 mm long; disk densely pubescent at margin with straight hairs, distinct free margin 0.2-0.5 mm long; style 25.5-42 mm long, exserted about as far as stamens. Fruit moderately to densely reddish-lepidote especially on body, 2.2-3.9 x 1.3-2.8 cm, oblong to elliptic or broadly elliptic, apex rounded or truncate to deeply retuse and scarcely apiculate, base rounded or truncate to retuse, very distinct narrow pseudostipe 0.08-0.3 cm long, wings 4, 0.6-1.1 cm wide. Scales ca. 70-250(-275) μm diam., marginal cells ca. 35-70.

Distribution: Centred on the Guianas and lower Amazon river-basin, extending more sparsely to W, N and S; in flooded, seasonally inundated or non-inundated deciduous or evergreen forest, often along rivers, on clay or sand, mainly in the lowlands; about 117 specimens seen from the Guianas (GU: 49; SU: 35; FG:33).

Selected specimens: Guyana: Berbice-Corentyne Region, banks of Corentyne R. above Baba-Grant Sawmill, above Cow Falls, McDowell & Gopaul 2302 (F, LTR, US); Kanuku Mts., Rupununi R., Puwib R., Jansen-Jacobs *et al.* 289 (CAY, LTR, U). Suriname: Wanica, Pikipada, Sauvain 735 (CAY, LTR, U); Coppename R., near Kaaimanston, Lanjouw 701 (BM, K, U). French Guiana: riverbank near Petit-Saut on Sinnamary R., Prévost 1312 (fl buds), 1313 (fr) (CAY, LTR); confluence of Litany and Waamahpann Rs., Tumuc-Humuc Mts., de Granville 12581 (CAY, LTR).

Vernacular names: Guyana: firebrush, kuyari engayi, monkey comb, monkey toothbrush, macaw comb. Suriname: arimaka, baredaballi, bosowi, bosroe, jalimana, yariman, keskesbosro, keskeskankan, maribena, sekema, vreemoessoe-noto. French Guiana: boso-oui, chigouma (fide Aublet 1775), pékéia, pumalipe.
All species of section **Combretum** have 'bottle-brush' inflorescences and are known very widely in various languages as monkey's toothbrush or monkey's comb.

Phenology: Flowering mostly August to May; fruiting mostly April to October.

9. **Combretum spinosum** Bonpl. in Humb. & Bonpl., Pl. Aequinoct. 2: 161. 1817. Type: Venezuela, Bolívar, Orinoco R., Angostura [Ciudad Bolívar], Humboldt & Bonpland 1118 (holotype P-B).

Woody liana to 3 m or stems trailing along ground; stem with strong infra-axillary spines up to 3 cm; 'combretaceous hairs' and stalked glands present. Leaves opposite or more or less so, chartaceous to subcoriaceous, elliptic to obovate-elliptic or narrowly so, 1.5-12.5 x 1.1-5.5 cm, apex rounded to acuminate, base cuneate to rounded, usually densely minutely verruculose along all veins, especially minor veinlets below, often hairless but sometimes sparsely pubescent especially on midvein below, with stalked glands rather frequent on veins below and sparse above but very small and inconspicuous; venation usually eucamptodromous, sometimes eucamptodromous-brochidodromous, secondary veins 4-9 pairs; petiole 0.2-1 cm long, glabrous to pubescent, with sparse stalked glands, often disarticulating at varying distances above base which remains to become spine next season. Inflorescence simple or branched, in leaf-axils, often forming a raceme or sparse panicle of spikes up to 16 cm long, each spike 1-11.5 cm long, slender, its axis sparsely to densely pubescent and with sparse to frequent stalked glands. Flowers 5-merous, 3.2-4.4 mm long; lower hypanthium 1.5-1.8 mm long, without pedicel-like region, usually slightly narrowed to neck, sparsely to densely appressed-pubescent, stalked glands absent to frequent, upper hypanthium campanulate to infundibuliform, 1.5-2.6 x 2-2.5 mm, more or less hairless outside, without or with very sparse stalked glands, sparsely pubescent inside; calyx lobes erect, triangular, 0.6-1 mm, apex subacute to obtuse; petals 5, patent to reflexed at anthesis, obovate to narrowly so, 1.5-1.6 x 0.8 mm, exserted, apex rounded or sometimes retuse, pubescent outside and on margins; stamens 10, well exserted, filaments 3.6-4.8 mm long; disk small but thick, closely applied to style base, pubescent at margin, free portion ca. 0.2 mm long; style 5.2-6.4 mm long, exserted about as far as filaments, sparsely pilose; ovules 5-8. Fruit glabrous and with very sparse stalked glands, 1-1.4 x 0.4-0.5 cm, linear to narrowly oblong in side view, sometimes wider near base, apex rounded to obtuse, base retuse without pseudostipe, ridges 5, strong, narrow, but without true wings, dark brown when dried. Stalked glands 35-80 µm, mostly appressed to organ, stalk uniseriate, ca. 5 cells long, obovoid head 25-40 x 15-25 µm.

Distribution: Apparently confined to Venezuela, but also very scattered and rare in the West Indies from Cuba to Trinidad; in dry deciduous or riverside rain forest; occurs close to Guyana in Venezuela (Bolívar): Reserva Forestal Imataca, Cuyuni R., between Isla Anacoco and Botanamo R., Stergios *et al.* 5272, 5294 (LTR, PORT).

3. **CONOCARPUS** L., Sp. Pl. 176. 1753.
Type: C. erectus L.

Mangrove-like shrubs or trees, without pneumatophores but sometimes with stilt-roots, not spiny; only 'combretaceous hairs' present. Leaves alternate, with conspicuous domatia (bowl-shaped pits) in secondary vein-axils; with petiolar glands. Inflorescences axillary or terminal racemes or panicles of compact, more or less globose heads, mainly leafless; bracts very small, withered by fruiting. Flowers possibly functionally dioecious but with a range of development of male and female organs, actinomorphic, sessile, 5-merous; upper hypanthium campanulate; calyx lobes 5; petals 0; stamens (5-)10, exserted, anthers versatile; style free, glabrous. Fruits small, flattened 2-winged nuts, upper hypanthium and calyx persistent, densely packed into cone-like heads which shatter to release individual fruits.

Distribution: A genus of 2 species: *C. erectus* in tropical America and tropical W Africa, and *C. lancifolius* Engl. & Diels, a non-mangrove shrub or tree of inland wet sandy ground in NE Africa and Arabia.

1. **Conocarpus erectus** L., Sp. Pl. 176. 1753. Type: [icon] Sloane, Voy. Jamaica t. 161, f. 2. 1725 (lectotype, designated by Wijnands, Bot. Commelins 66. 1983); typotype: Herb. Sloane 5: 63 (designated by C.E. Jarvis *et al.*, Regnum Veg. 127: 37. 1991). – Fig. 8

Evergreen mangrove-like shrub or tree 2-20 m, sometimes with stilt-roots. Leaves coriaceous, narrowly elliptic or sometimes elliptic, 3-12 x 0.6-3.5 cm (incl. petiole), apex tapering narrowly acute to acuminate, base gradually decurrent to narrowly acute, usually glabrous or nearly so, except often sericeous on petiole and at least basal part of midvein below; domatia conspicuous bowl-shaped pits; venation brochidodromous, secondary veins 4-7 pairs; petiole 0-3 mm long, biglandular on petiole or at base of leaf. Inflorescence axillary or terminal raceme or sparse panicle of ± globose flower-heads, up to ca. 15 cm long but usually much less, sometimes with leaves at lower nodes; possibly functionally dioecious, with range of morphological development of male and female organs, often apparently male and bisexual flowers in same head; peduncles densely silvery-sericeous; flower-heads 3-5 mm diam.. Flowers 2.5-2.8 mm long (incl. calyx); lower hypanthium 0.7-1.5 x 1.4-1.7 mm in bisexual flowers, densely appressed-pubescent, upper hypanthium cupuliform, 0.8-1.2 x 1.2-1.3 mm (incl. calyx), appressed-pubescent; calyx lobes erect to incurved, 0.3-0.5 mm long; stamens (5-)10, exserted up to 2 mm in male flowers,

Fig. 8. *Conocarpus erectus* L.: A, female plant, flowering shoot; B, inflorescence; C-D, flowers from two faces; E, male plant, flowering shoot; F, flower with front of perianth removed; G, half-flower; H, domatium in secondary vein axil, lower surface of leaf; I, fruiting shoot; J, infructescence (A-D, Balick *et al.* 1949, Belize; E-H, Vincelli 579, Honduras; I-J, Knapp *et al.* 3400, Panama). Drawing by Rosemary Wise.

otherwise variously shorter or aborted; disc pilose; style 0.5-1.7 mm long, often bent or with S-shaped kink, slightly exserted, glabrous. Fruits packed densely into globose to ellipsoid heads of 5-15 x 7-13 mm, only upper part of fruit exposed; fruit puberulous on convex face, glabrous on concave face, 3.7-4 x 4-4.2 mm, flattened, broadly obovate to orbicular in side view, bearing old upper hypanthium until at least maturity, ± 2-winged, wings ca. 0.5-1.5 mm wide.

Distribution: Atlantic and Pacific coasts of tropical America from Baja California Norte (ca. 29° N) and Bermuda (32° 20' N) throughout the West Indies to Rio de Janeiro, Brazil (ca. 23° S) and extreme N Peru (ca. 3° 30' S), incl. Galápagos, in W Africa from Senegal (ca. 15° N) to Angola (ca. 7° S); at landward fringe of mangrove swamps, sometimes codominant with *Laguncularia* and *Rhizophora*, usually within reach of the highest tides, also found on sandy shores, exposed coral limestone, salt-marshes, dry or inundated savanna, and primary or secondary woodland, showing somewhat weedy tendencies and being tolerant of saline and fresh water, and dry soils; usually at sea-level, but commonly up to 20 m elev.; about 6 specimens from the Guianas (GU: 4; SU: 1; FG: 2).

Selected specimens: Guyana: Berbice, Waterloo Station, Jenman 4997 (K); Demerara, Hope Beach, on white sand, de Granville & Poncy 11702 (K); Essequibo R., Essequibo Isl., Naamryk Canal just W of Lookout, at head of canal facing, Pipoly & Samuels 11751 (LTR, US). Suriname: Nickerie, NE of Prodobong, E of Nickerie, Geyskes s.n. (K, U). French Guiana: Cr. Caswine, Benoist 780 (P).

Vernacular names: Guyana: buttonbush, button mangrove.

Phenology: Flowering and fruiting throughout most of year.

Note: Often considered a mangrove species, but it is better treated as mangrove-associate because of its lack of pneumatophores and vivipary. All Guianan material belongs to forma *erectus*, with glabrous or subglabrous leaves.

4. **LAGUNCULARIA** C.F. Gaertn., Suppl. Carp. 209. 1807.
Type: L. racemosa (L.) C.F. Gaertn. (Conocarpus racemosus L.)

Mangrove shrubs or trees, with often simple pneumatophores and stilt-roots, not spiny; only 'combretaceous hairs' present apart from glandular pits on leaves. Leaves opposite, scattered but mainly submarginal minute pits on both surfaces each with basal sessile gland, without domatia; with petiolar glands. Inflorescences usually leafless axillary or terminal spikes or racemes of spikes; bracts very small, caducous. Flowers mostly bisexual but often some unisexual (mostly male) ones and sometimes (? usually) dioecious, actinomorphic, sessile, 5-merous; 2 lateral prophylls (bracteoles) fused to lower hypanthium, upper hypanthium short, cupuliform; calyx lobes 5; petals 5, slightly exceeding calyx lobes; stamens 10, included or ± so, anthers versatile; style free, glabrous. Fruits slightly compressed longitudinally ridged nuts, upper hypanthium and calyx persistent, and 2 prophylls still visible.

Distribution: A single species on the Pacific and Atlantic coasts of America and the Atlantic coast of Africa.

LITERATURE

Valente, M. da C., N. Marquete Ferreira da Silva & D.J. Guimarães. 1994. Morfologia e anatomia do fruto de Laguncularia racemosa (L.) Gaertn. f. (Combretaceae). Arch. Jard. Bot. Rio de Janeiro 32: 39-50.

1. **Laguncularia racemosa** (L.) C.F. Gaertn., Suppl. Carp. 209. 1807. – *Conocarpus racemosus* L., Syst. Nat. ed. 10. 930. 1759. Type: Herb. Linnaeus No. 237.2 (lectotype LINN) (designated by Stace in Bornstein 1989: 459). – Fig. 9

Laguncularia obovata Miq., Linnaea 18: 752. 1844. Type: Suriname, Hostmann 665 (lectotype BM, isolectotypes CGE, G, K) (designated by Stace 2007b: 10).

Evergreen mangrove shrub or tree 0.5-12 m, often simple 'peg-root' pneumatophores. Leaves coriaceous or somewhat succulent, elliptic to oblong-elliptic, 3-12 x 2-6 cm, apex obtuse to rounded or sometimes retuse, base rounded to obtusely cuneate, glabrous when fully expanded but usually sericeous when very young; minute pits, each with basal gland, visible as pimples on both leaf surfaces, usually mostly submarginal; venation brochidodromous, secondary veins 9-18 pairs, not prominent; petiole 0.7-2(-2.5) cm long, usually sericeous at first, soon glabrous, with 2 sessile glands. Inflorescence 2-20 cm long, axillary or terminal simple spikes or racemes of spikes, variously with male, female or bisexual flowers, plants bisexual to dioecious; inflorescence axes and young stems usually densely sericeous; rhachis up to 18 cm long. Male flowers ca. 2 x 3-4(-5) mm, female/bisexual ones 3.5-6.5 x 3-4(-5) mm, gradually elongating in fruit; lower hypanthium obconical, 2-4 mm long

84

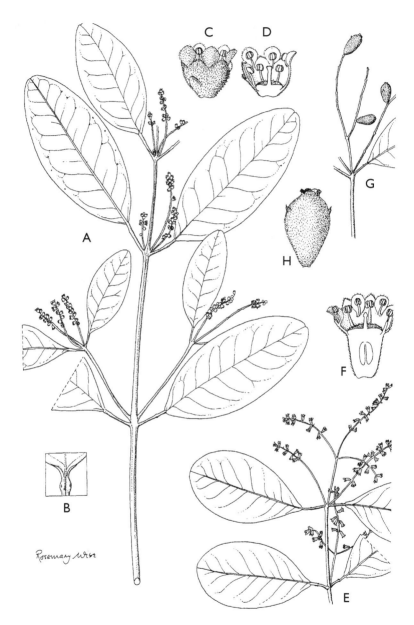

Fig. 9. *Laguncularia racemosa* (L.) C.F. Gaertn.: A, male plant, flowering shoot; B, apex of petiole; C, flower; D, half-flower;.E, female plant, flowering shoot; F, half-flower; G, infructescence, H, fruit (A-D, Hammel 3299, Panama; E-F, Jacquemin 2844, French Guiana; G-H, Callejas & Bornstein 11028, Colombia). Drawing by Rosemary Wise.

in bisexual/female flowers, densely fulvous-sericeous or rarely glabrous, 2 subapical prophylls (bracteoles) each resembling a calyx-lobe, upper hypanthium shallowly cupuliform, 1.5-2.5 mm long, incl. erect to incurved calyx lobes of ca. 1 mm long, densely fulvous-sericeous or rarely glabrous; petals broadly ovate, ca. 1.5 mm long, apex obtuse, pubescent outside; stamens 1-2 mm long, scarcely exserted; disk pubescent at outer edge; style 1-1.5 mm long, scarcely exserted, glabrous. Fruit densely fulvous-sericeous to -tomentose, rarely glabrous, 12-20 x 4-10 mm, obconical to fusiform, somewhat flattened laterally, retaining upper hypanthium, calyx and bracteoles but losing petals and stamens, longitudinally irregularly ridged, 2 narrow longitudinal wings up to 1.5 mm wide running down from bracteole midveins.

Distribution: Atlantic and Pacific coasts of tropical America from Baja California Sur (27° N), Florida (28° 30' N) and Bahamas throughout the West Indies to Santa Catarina (ca. 28° 30' S) and extreme N Peru (ca. 3° 30' S), incl. Galápagos, in W Africa from Senegal (ca. 12° 30' N) to Angola (ca. 9° S); characteristic mangrove species in tropical America, often occurring with *Rhizophora, Avicennia,* etc., but sometimes in pure stands, in mangrove swamps, usually on the landward side, in standing water or daily submerged, in other strand vegetation, and also in coastal scrub or disturbed ground only irregularly flooded, at 0-10 m elev.; about 13 specimens from the Guianas (GU: 3; SU: 3; FG: 7).

Selected specimens: Guyana: Demerera-Mahaica region, Atlantic coast, W of Mahaica R., Hoffman *et al.* 732 (K, LTR, U, US); Barima-Waini Region, 2 miles NW of Shell Beach camp, towards Waini turtle nesting area, Hollowell 206 (LTR, US). Suriname: Marowijne, on beach at Wiawia Bank, Grote Zwiebelzwamp, Lanjouw & Lindeman 1059 (K, U); Paramaribo, at edge of forest, Kappler s.n. (G). French Guiana: Île de Cayenne, pied de la colline de Burda, chenal en bord de mer, Sabatier 2259 (CAY, LTR); Beira do Approuague R., Black *et al.* 54-17593 (BM).

Vernacular names: general: white mangrove, mangle blanco. Guyana: furilia. Suriname: ankira, akira, furilia. French Guiana: pelétuvier gris.

Phenology: Flowering and fruiting throughout most of year.

Note: Produces distinctive pneumatophores, but the fruit is only slightly viviparous, the embryo rupturing the testa and occasionally the radicle protruding from the fruit while still on the tree. All Guianan material belongs to var. *racemosa,* with inflorescence axes, fruits and flowers, and young stems, leaves and petioles densely fulvous-sericeous.

5. TERMINALIA L., Syst. Nat. ed. 12. 2: 674. 1767, nom. cons.
Type: T. catappa L.

Bucida L., Syst. Nat. ed. 10. 1025. 1759, nom. cons.; it is a nom. rej., however, against Terminalia.
Type: B. buceras L. (= Terminalia buceras (L.) C. Wright)
Tanibouca Aubl., Hist. Pl. Guiane 448. 1775.
Type: T. guianensis Aubl. (= Terminalia dichotoma G. Mey.)

Trees or shrubs, from ca. 0.5 m to 60(-70) m, taller ones often with buttresses, rarely spiny; only 'combretaceous hairs' present. Leaves spirally arranged, usually clustered at branchlet tips, sometimes with pocket-shaped or bowl-shaped domatia in secondary vein-axils; usually with petiolar glands. Inflorescences axillary lax to congested simple leafless spikes, spikes often clustered at branchlet-ends; bracts very small and caducous. Flowers bisexual or andromonoecious, actinomorphic, sessile, (4-)5-merous; lower hypanthium extended into a usually short 'neck'; upper hypanthium cupuliform to campanulate, deciduous before fruiting or sometimes persistent; calyx lobes (4-)5; petals 0; stamens (8 or)10, usually well exserted, anthers versatile; disk glabrous to densely pubescent; style free, usually exserted, glabrous or pubescent, baselly usually pubescent, glabrous towards apex. Fruit 2-5-winged or -ridged or terete, flattened to actinomorphic, usually dry or spongy, sometimes slightly succulent.

Distribution: At least 200 species throughout the tropics, extending to subtropical regions in S America (to 37° 15' S) and N America (to 26° 40' N), the greatest number of species and variation occur in E tropical Asia; only 34 species occur in America; 7 species in the Guianas.

Notes: A world-wide sectional classification has not been fully worked out; at least 25 sections are currently recognised but several more should be described, especially in Asia; 12 of them occur in America and 7 sections in the Guianas. The sections are not used here.
In the Flora of Suriname, *Terminalia buceras* (L.) C. Wright was mentioned to occur in Suriname and Guyana, in Exell 1935: 174 as *Buceras bucida* Crantz, in Görts-van Rijn 1986: 354 corrected to *Bucida buceras* L. The specimen Splitgerber 354, from the sea-coast at Suriname, collected in flower in December 1837 and seen by Pulle for his Enumeration of vascular plants of Surinam (1906: 342), could not be traced by Exell (1935) and the present author. It is not in BM, BR, CGE, K, GOET, L or U. Pulle also mentioned its occurrence in Guyana, which I have again been unable to confirm.

LITERATURE

Flores, E.M. 1994. Roble coral, Bullywood (Terminalia amazonia). Árboles Semillas Neotropico 3(1): 55-86.

KEY TO THE SPECIES

1   Leaves cordate at base; fruits drupe-like, with lateral ridges but without wings, slightly compressed . . . . 2. *T. catappa*
    Leaves cuneate to attenuate at base; fruits distinctly winged, sometimes narrowly so but then fruits strongly flattened 2

2   Mature fruits with 3 zones across width (body; adjacent part of wing markedly thickened with spongy material and scarcely thinner than body; marginal thin part of wing) 3
    Mature fruits with 2 zones across width (body; thin wing) 4

3   Leaves up to 22 cm long, secondary veins distant; fruit usually slightly longer than wide, glabrous . . . . 3. *T. dichotoma*
    Leaves up to 14 cm long, secondary veins moderately spaced; fruit usually slightly wider than long, densely pubescent when young . . . . 5. *T. lucida*

4   Fruits 1-3.2 cm long, with wings 0.8-3.5 cm wide; style densely pubescent at least basally 5
    Fruits 0.5-1.2(-1.5) cm long, with wings 0.2-1 cm wide; style glabrous, or very sparsely pubescent near base 6

5   Inflorescence 1.5-5.5 cm long, with rhachis 0.8-3.5 cm long; disk sparsely pilose . . . . 4. *T. guyanensis*
    Inflorescence 9-17 cm long, with rhachis 8-14 cm long; disk pubescent . .
    . . . . 6. *T. oblonga*

6   Leaves chartaceous-subcoriaceous, domatia present, tertiary veins closely and regularly percurrent; fruit wider than long, with 2-5 wings, if more than 2 then 2 much wider than the others . . . . 1. *T. amazonia*
    Leaves strongly coriaceous, domatia absent, tertiary veins randomly reticulate; fruit longer than wide, with (4-)5 equal wings . . . . 7. *T. quintalata*

1. **Terminalia amazonia** (J.F. Gmel.) Exell in Pulle, Fl. Suriname 3(1): 173. 1935. – *Chuncoa amazonia* J.F. Gmel., Syst. Nat. 2: 702. 1791. Type: not designated.

Bucida angustifolia DC., Prodr. 3: 10. 1828. – *Bucida buceras* L. var. *angustifolia* (DC.) Eichler in Mart., Fl. Bras. 14(2): 95. 1867. Type: French Guiana, Cayenne, Perrottet s.n. (lectotype G) (designated by Stace 2007b: 39).

88

Terminalia odontoptera Van Heurck & Müll. Arg. in Van Heurck, Observ. Bot. 2: 217. 1871. Type: French Guiana, Insula Cayenne, Patris s.n. (lectotype G, isolectotype AWH, not seen) (designated by Stace 2007b: 39).

Deciduous tree (4-)10-60(-70) m, with plank buttresses, sometimes branching, up to 3 m high and up to 2 m from trunk. Leaves chartaceous to subcoriaceous, usually obovate, less often obovate-oblong or -elliptic or narrowly so, (2-)4-10(-15) x 2-5(-6.5) cm, apex rounded to acute or gradually or abruptly acuminate, base cuneate to attenuate-cuneate, glabrous to sparsely pubescent above, subglabrous to pubescent except usually pubescent on main veins below; domatia pocket-shaped or bowl-shaped; venation eucamptodromous or eucamptodromous-brochidodromous, midvein moderate, prominent, secondary veins 3-5 pairs, distant, originating at moderately or sometimes widely acute angles, curved, prominent, intersecondary veins usually absent, tertiary veins closely and very regularly and conspicuously percurrent; higher order veins occasionally distinct; areolation imperfect or incomplete, slightly prominent petiole 0.2-1.5 cm long, eglandular or biglandular (sometimes more). Inflorescence 5-24 cm long, simple, all flowers bisexual or some to many with reduced female parts; peduncle 1-3 cm long, appressed-pubescent; rhachis 4-21 cm long, appressed-pubescent. Flowers 5-merous, 2.5-4 x 2.5-3.5 mm; lower hypanthium 1.5-2 mm long, densely appressed- or sometimes patent-pubescent, upper hypanthium shallowly campanulate to shallowly infundibuliform, 0.5-1 mm long, appressed-pubescent or sparsely so; calyx lobes suberect to patent, 0.3-0.8 mm long, appressed-pubescent or sparsely so; stamens 2-4 mm long; disk subglabrous to pilose; style 2.5-3.5 mm long, glabrous. Infructescence with numerous fruits disposed along whole length of rhachis; fruit usually straw-coloured, pubescent to sparsely pubescent, 0.5-0.7(-1) x 0.8-1.8(-2.2) cm, variously flattened, transversely oblong in side view, apex truncate but shortly beaked, base truncate or slightly retuse but sometimes very shortly pseudostipitate, wings 2-5, flexible, unequal, 2 lateral ones 0.3-1 cm wide, rounded or narrowly rounded laterally, the other 3 varying from strong ridges to wings up to 2 mm wide, body 0.15-0.35 cm wide, bulging and keeled to winged on both faces.

D i s t r i b u t i o n : Widespread from Mexico (Veracruz) to central Bolivia and central Brazil (Pará, Tocantins); tall often emergent forest tree found in many situations: primary or secondary forest, high or low forest, dry forest on hills, dense swamp forest, evergreen rain forest, gallery forest along rivers, river terraces, semi-deciduous forest, among mangroves, mostly lowland, but up to 700 m elev.; about 82 specimens from the Guianas (GU: 18; SU: 27; FG: 37).

Selected specimens: Guyana: Berbice, Warunana Cr., left bank of Ituni R., Berbice R., Hohenkerk 136B (K); E Demerara Region, Yarowkabra settlement and Forestry Station, Timehri-Linden Highway, Pipoly *et al.* 7438, 7459 (FDG, LTR, NY, U, US); Takutu-Upper Essequibo Region, S Pakaraima Mts., 17 km NW of Karasabai, mouth of Tipuru R. at Ireng R., Hoffman *et al.* 1027 (LTR, U, US). Suriname: Marowijne, Tamarin, Kock s.n. (U); Lely Mts., 175 km SSE of Paramaribo, Mori & Bolten 8450 (LTR, NY); Saramacca R., W of Poika, Schulz 7628 (U); Wanica, Banafokondre, Suriname R., Sauvain 157 (CAY, LTR). French Guiana: Piste de Sainte-Elie, interfleuve Sinnamary/Counamama, Sabatier & Prévost 4188 (CAY, LTR); Station des Nouragues, Arataye R. basin, Sabatier 3515 (CAY, LTR).

Vernacular names: Guyana: fukadi (this name is used for several species of *Terminalia* and *Buchenavia*). Suriname: anagosuti, bosamandel, foekadie, gemberhout, gindya-udu, ginja hoedoe, itju tano kawai, itoetanoe kwai, kalebashout, katurimja, komanti kwatii, kwai, kuware, sansan-udu, saraija. French Guiana: katuma.

Phenology: Flowering and fruiting most months of the year.

Use: An important timber known as Nargusta or Roble Coral in the international trade.

2. **Terminalia catappa** L., Syst. Nat. ed. 12. 2: 674. 1767 & Mant. Pl. [1]: 128. 1767. – *Myrobalanus catappa* (L.) Kuntze, Revis. Gen. Pl. 1: 237. 1891. Type: Herb. Linnaeus No. 1222.1 (lectotype LINN) (designated by Byrnes, Contr. Queensland Herb. 20: 38. 1977).

Terminalia paraensis Mart., Flora 24(2, Beibl.): 24. 1841. Type: Brazil, Pará, Martius s.n. (holotype M).

Evergreen to briefly deciduous tree 2-35 m. Leaves chartaceous, obovate to broadly so or rarely elliptic-obovate, (8-)12-30(-38) x (5-)9-15(-22) cm, apex rounded to shortly acuminate, base tapering to usually cordate to subcordate (rarely rounded, subtruncate or cuneate), glabrous above, glabrous to appressed-pubescent below; domatia bowl-shaped, always present; venation eucamptodromous-brochidodromous, secondary veins 6-12 pairs, moderately spaced to distant, originating at moderately to widely acute angles, curved distally, prominent, intersecondary veins present; tertiary veins usually irregularly percurrent, often alternate and oblique; quaternary veins sometimes conspicuous; areolation randomly reticulate, imperfect or incomplete; petiole 0.5-2.5 cm long, pubescent, usually biglandular. Inflorescence (8-)13-30 cm long, simple, andromonoecious, with bisexual flowers few and near base; peduncle 3-5.5 cm long, glabrous to sparsely pubescent; rhachis (5-)10-27 cm long, pubescent. Flowers 5-merous, 3-5 x 4-7 mm (male) or 6-10 x 4-7 mm (bisexual); lower hypanthium 3-7 mm long in bisexual flowers, appressed-pubescent, usually densely so near base and sparsely so near apex, upper hypanthium cupuliform or campanulate, 1-2 mm long, sparsely pubescent; calyx lobes erect to patent or slightly recurved when at full anthesis, 1-1.5 mm long, nearly glabrous; stamens 2-4 mm long; disk villous; style 3-3.5 mm long, glabrous. Infructescence with few fruits near base of rhachis; fruit drupaceous but rather fibrous, glabrous, (3.5-)4-8 x 3-5.5 cm, ovoid to ellipsoid, slightly compressed, apex acute to acuminate or stoutly beaked, base rounded to broadly cuneate, obscure to conspicuous ridge or wing up to 6 mm wide along full length on each lateral edge.

Distribution: Native in tropical Asia, probably from India to the extreme SE, and in Polynesia and NE Australia, naturalised throughout the Neotropics; ornamental tree introduced into America certainly by the early 1800s, sometimes thoroughly naturalised on marginal open ground, especially close to the sea where it may form a zone behind the mangroves, it tolerates salinity, mostly at low elevation, but up to at least 1470 m; most American herbarium material does not make it clear whether the specimen came from naturalised (wild) or cultivated trees; about 20 specimens from the Guianas (GU: 5; SU: 6; FG: 9).

Selected specimens: Guyana: Barima-Waini Region, Shell Beach camp, 5 miles SE of Waini Point, Hollowell et al. 264 (LTR, US). Suriname: Marowijne, near Albina, Marowijne R., Versteeg 553 (U). French Guiana: Île de Grand Connétable, Lescure 272 (CAY, K, U).

Vernacular names: Various references to the almond-like kernels. Widespread: almendro, indian almond. Suriname: amandelboom, mantara.

Phenology: Flowers and fruits are found throughout most of the year, even in one locality.

Uses: Widely grown as an ornamental and for shade. The fruits have a spongy mesocarp and are well adapted for water dispersal. The kernels (seeds) are eaten like almonds.

3. **Terminalia dichotoma** G. Mey., Prim. Fl. Esseq. 177. 1818. – *Terminalia latifolia* Sw. var. *dichotoma* (G. Mey.) DC., Prodr. 3: 12. 1828. Type: Guyana, Essequibo R., Arowabish Isl., G. Meyer 113 (holotype GOET).

Tanibouca guianensis Aubl., Hist. Pl. Guiane 1: 448. 1775, non *Terminalia guyanensis* Eichler 1867. – *Myrobalanus guianensis* (Aubl.) Kuntze, Revis. Gen. Pl. 1: 237. 1891. Type: French Guiana, Aublet s.n. (lectotype BM) (designated by Stace 2007b: 33).

Tree 3-60 m (? deciduous or evergreen), sometimes with large plank buttresses. Leaves chartaceous to subcoriaceous, obovate or obovate-oblong to oblanceolate, 9-22 x 4-10 cm, apex acuminate, base narrowly cuneate, pubescent when young then glabrous or nearly so when mature above, densely pubescent when young then subglabrous to sparsely pubescent when mature below; domatia absent; venation eucamptodromous or eucamptodromous-brochidodromous, midvein moderate, prominent, secondary veins 5-8(-10) pairs, distant, originating at moderately to widely acute angles, curved, prominent, intersecondary veins present; tertiary veins irregularly percurrent; higher order veins usually distinct; areolation large, well-developed to imperfect, slightly prominent; petiole 1.5-4.5 cm long, pubescent when young, becoming glabrous or ± so, usually biglandular. Inflorescence 7-14 cm long, simple, all flowers bisexual; peduncle 1.5-3 cm long, glabrous to sparsely pubescent; rhachis 5.5-11 cm long, glabrous to sparsely pubescent. Flowers 5-merous, 4-6 x 4-5.5 mm; lower hypanthium 2-3 mm long, very sparsely appressed-pubescent to tomentose with rufous hairs, upper hypanthium campanulate, 1.5-2 mm long, glabrous to sparsely pubescent; calyx lobes revolute, 1.5-2 mm long, glabrous to sparsely pubescent; stamens 4-6 mm long; disk villous; style 5-6 mm long, villous in basal half. Infructescence with usually few fruits, disposed along whole length or mostly near apex of rhachis; fruits glabrous, 2.5-4.5 x 2-4.5 cm, strongly flattened, suborbicular to very broadly elliptic or obtriangular in side view, apex rounded, emarginate or apiculate, base rounded to cuneate (not pseudostipitate), wings 2, stiff, equal, 7-15 mm wide, markedly thickened with spongy material adjacent to body and for most way to margin (thin part of wing only ca. 4-6 mm wide), rounded to strongly angled laterally (the angle nearer the fruit apex), body 0.6-1.2 cm wide but appearing much wider due to thickened base of wings, often 1-2-ridged on 1 face and slightly grooved on other.

Distribution: Widespread in S America, from Colombia and Venezuela to Peru and Brazil (Bahia), mostly sublittoral in the E, but scattered along the Amazon river-basin to Ecuador and Peru, much sparser in the W, especially characteristic of the Guianas; mainly a tree of riverine or inundated often primary forest on alluvial or sandy soils, up to 250 m elev.; about 69 specimens from the Guianas (GU: 26; SU: 13; FG: 30).

Selected specimens: Guyana: Berbice, Coomaka, Persaud 187 (F, K, NY, R, US); Mazaruni-Potaro Distr., Thomas Isl., Essequibo R. near Bartica, Mori *et al.* 8153 (LTR, NY, U, VEN). Suriname: Marowijne R., Tapanahoni R., Versteeg 647 (P, U); Sipaliwini, Kuruni R., 30 km E of conflluence with Corantyne R., N side of Kuruni Isl., Evans *et al.* 1910 (K, U); Saramacca, Coppename R., near Raleigh Falls, Lanjouw 815 (K, NY, R, S, U). French Guiana: Approuague R., at Petit Saut Couota, Oldeman B-1926 (CAY, U); Région de Zidockville, Upper Oyapock R., Prévost & Grenand 977 (CAY, LTR).

Vernacular names: Guyana: coffee mortar, fukadi, maharu, naharu, swamp fukadi. Suriname: alaso-abo, bosamandel, boskalebas, foekadi djamaro, fukadi, gindaya-udu, kalebashout, karalawai jakoenepele, kararawa akunepere, sansan-udu, zwamp bosamandel. French Guiana: alalamunuwi.

Phenology: Flowering June to October; fruiting October to December.

Use: A hard timber used locally for house construction.

4. **Terminalia guyanensis** Eichler in Mart., Fl. Bras. 14(2): 88. 1867. Type: French Guiana, Poiteau s.n. (holotype B destroyed, photo at FI, isotypes K, P, U, W). – Fig. 10

Deciduous tree 6-40 m, with plank buttresses to 3 m high, 1.7 m wide, 0.15 m thick. Leaves chartaceous, elliptic-oblong to obovate or narrowly so, 4-11 x 2-4(-5.5) cm, apex long-acuminate,base cuneate or attenuate-cuneate, sericeous when young, glabrous or nearly so except sparsely pubescent on main veins below at maturity; domatia absent; venation brochidodromous, midvein moderate, prominent, secondary veins (5-)7-12 pairs, moderately spaced to distant, diverging at widely acute angles, curved, prominent, intersecondary veins often present, tertiary veins randomly reticulate or sometimes weakly percurrent; higher order veins not distinct; areolation large, imperfect, scarcely prominent; petiole 1.2-3.5 cm long, usually glabrous, biglandular. Inflorescence 1.5-5.5 cm long, simple, all flowers bisexual; peduncle 0.5-2.8 cm long, pubescent; rhachis 0.8-3.5 cm long, subcapitate to moderately elongated, densely pubescent at flowering. Flowers 5-merous, 3-4.6 x 2-3.8 mm; lower hypanthium 1.5-2.2 mm long, pubescent to densely so, upper hypanthium deeply cupuliform to deeply campanulate, 0.8-1.5 mm long, pubescent to sparsely so; calyx lobes revolute to erect, 0.7-1 mm long, pubescent to sparsely so; stamens 3-5 mm long; disk sparsely pilose; style 3-4 mm long, pilose basally or

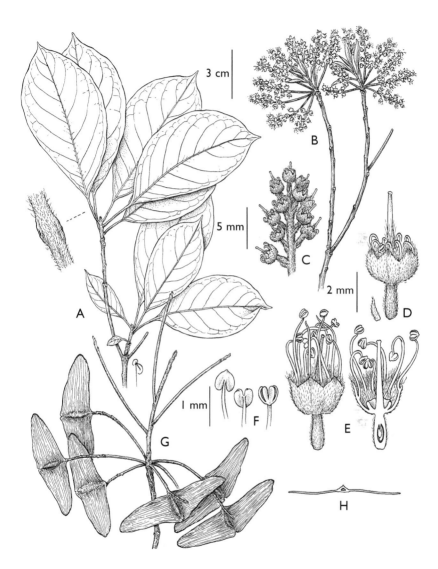

Fig. 10. *Terminalia guyanensis* Eichler: A, leafy shoot, with enlarged petiole showing glands; B, flowering shoot; C, apical part of spike; D, flower and bract at female stage; E, flower and half-flower at male stage; F, stamens in 3 views; G, fruiting shoot; H, fruit in transverse section (A, Mori *et al*. 23905; B-F, Mori & Gracie 18653; G-H, Mori & Boom 15121). Drawing by Bobbi Angell, reprinted from Mori *et al*., 2002, with permission of B. Angell, S. Mori and New York Botanical Garden.

almost to apex. Infructescence with few fruits crowded on short rhachis; fruit glabrous and usually shiny, 1.6-3.2 x 3.3-7.8 cm, flattened, transversely oblong in side view, apex rounded to emarginate, base obtuse (often with concave basiscopic wing margin), truncate or slightly emarginate, wings 2, fairly stiff, equal, 1.7-3.5 cm wide, rounded to very narrowly rounded laterally, up to 4 cm long at longest point, body 0.4-0.7 cm wide, usually bulging on both faces.

D i s t r i b u t i o n : W Venezuela to French Guiana; tropical moist forest, never far inland, on clay soils, at 200-600 m elev. in the Guianas; about 16 specimens from the Guianas (GU: 3; SU: 2; FG: 11).

S e l e c t e d   s p e c i m e n s : Guyana: Rupununi Distr., E Kanuku Mts., NE of Warimure, Jansen-Jacobs et al. 2191 (K, LTR, U). Suriname: Nickerie, 22 km SW of Avanavero Dam site, Kabelebo Dam Project, Heyde & Lindeman 89 (F, K, MO); Nassau Mts., Lanjouw & Lindeman 2853 (U). French Guiana: St. Georges-Regina, between Pte. 30.6 & 31.85, Grenand 3065 (CAY, LTR); Saül, Mts. La Fumée, at entrance to La Fumée Trail on W side, Mori et al. 15121 (CAY, LTR, NY).

V e r n a c u l a r   n a m e s : Guyana: fukadi. Suriname: guaybon, sansan-hoedoe, sansan-udu.

P h e n o l o g y : Flowering October to March; fruiting October to June.

U s e : Timber said to be very hard.

5. **Terminalia lucida** Hoffmanns. ex Mart. & Zucc., Flora 7 (1, Beil.): 130. 1824. – *Myrobalanus lucida* (Hoffmanns. ex Mart. & Zucc.) Kuntze, Revis. Gen. Pl. 1: 237. 1891. Type: Brazil, Pará, Siber, von Hoffmannsegg s.n. (lectotype BR, isolectotype BM) (designated by Stace, Flora Mesoamericana, in press).

Terminalia eriantha Benth., Hooker's J. Bot. Kew Gard. Misc. 2: 240. 1850. Type: Brazil, Pará, Caripi, Spruce 166 (holotype K, isotype P).

Briefly deciduous tree 5-20 m. Leaves thickly chartaceous to coriaceous, oblong- or elliptic-obovate or narrowly so, 4.5-14 x 2-7.5 cm, apex subacute to shortly acuminate,base cuneate to attenuate, glabrous except sometimes appressed-pubescent on veins; domatia absent; venation brochidodromous or eucamptodromous-brochidodromous, midvein moderate, prominent, secondary veins 5-9 pairs, moderately spaced, originating at moderately acute angles, slightly curved distally, scarcely prominent, intersecondary veins rarely present, tertiary veins usually randomly reticulate, sometimes weakly percurrent; higher order veins usually well-developed; areolation close, usually well developed, not to slightly prominent; petiole 0.7-2 cm long, glabrous to pubescent, eglandular. Inflorescence 7.5-12 cm long, simple, all flowers bisexual; peduncle 1.5-2 cm long, pubescent; rhachis 6-10 cm long, densely pubescent. Flowers 5-merous, 3-4 x 2.5-3.5 mm; lower hypanthium 1-2 mm long, rufous- to white-tomentose, upper hypanthium cupuliform to campanulate, 1.5-2 mm, pubescent to tomentose; calyx lobes usually revolute, 0.6-1 mm long, pubescent; stamens 2.5-4 mm long; disk villous; style 2.8-4 mm long, villous except glabrous near apex. Infructescence with usually few fruits, disposed along whole length or mostly near apex of rhachis; fruit densely pubescent at first, becoming subglabrous at maturity, 1.7-3 x 2.5-3.8 cm, strongly flattened, rhombic or suborbicular to transversely broadly elliptic in side view, apex rounded or truncate to emarginate, base pseudostipitate, wings 2, stiff, equal, 0.9-1.4 cm wide, markedly thickened with spongy material adjacent to body, rounded to somewhat angled laterally, sometimes 1-2 extra narrower wings on more convex side of body, body 0.6-0.9 cm wide but appearing wider due to thickened base of wings, often slightly ridged on 1 face and slightly grooved on other.

Distribution: Guatemala to Panama, in S America from Colombia to Bahia in Brazil, at or near the coast except in E Brazil where it extends up to ca. 1200 km inland along the Tocantins, Paraná and Araguaia rivers, in W Africa from Guinea Bissau through Guinea to Sierra Leone; inundated lowland forest, riversides, river deltas and beaches, usually on sandy soils, often in a zone behind the mangroves just above high-water level; about 20 specimens from the Guianas (GU: 2; SU: 4; FG: 14).

Selected specimens: Guyana: Essequibo, Suddie, on seashore, Forest Dept. 5481 = Field no. 2688 (FDG, K). Suriname: Christiankondre, lower Marowijne R., Hekking 1077 (U); Wiawia Reserve, 3 km E of Motkreek, Sterringa 12480 (NY, U), 12527 (U). French Guiana: tree fallen on beach at Montjoly, near Pointe de Montravel, Cremers 9332 (BM, BR, CAY, K, LTR, U, US); Maroni, Sagot 1007 (BM, K, P).

Vernacular names: Guyana: fukadi (this name is used for several species of *Terminalia* and *Buchenavia*).

Phenology: Flowering and fruiting August to March.

6. **Terminalia oblonga** (Ruiz & Pav.) Steud., Nomencl. Bot. ed. 2. 2:
    668. 1841. – *Gimbernatia oblonga* Ruiz & Pav., Syst. Veg. Fl. Peruv.
    Chil. 275. 1798. – *Chuncoa oblonga* (Ruiz & Pav.) Pers., Syn. Pl. 1:
    486. 1805. – *Myrobalanus oblonga* (Ruiz & Pav.) Kuntze, Revis.
    Gen. Pl. 1: 237. 1891. Type: Peru, Huánuco, Pozuzo , Pavón s.n.
    (holotype MA, isotypes F, FI, MA, OXF).

    Terminalia obidensis Ducke, Arch. Jard. Bot. Rio de Janeiro 4: 147. 1925.
    Type: Brazil, Pará, Obidos, Ducke 17676 (lectotype MG, not seen,
    isolectotypes RB, S, U, US) (designated by Stace 2007b: 35).

Tree up to 50 m, probably briefly deciduous, with simple buttresses up
to 5 m high and 1.5 m wide. Leaves chartaceous, oblong- or elliptic-
obovate or -oblanceolate, 6-20(-25) x 4-7(-10) cm, apex acute to shortly
acuminate, base cuneate (usually narrowly so), sparsely pubescent above
and densely so below when young, glabrous or sparsely pubescent only
on main veins at maturity; domatia absent; venation brochidodromous to
eucamptodromous-brochididromous, midvein stout to moderate,
prominent, secondary veins 5-7 pairs, distant, originating at widely acute
angles, curved, prominent, inter-secondary veins usually present, tertiary
veins randomly reticulate or sometimes weakly percurrent; higher order
veins distinct or not; areolation well developed to imperfect, prominent
or not.; petiole 0.2-2(-5) cm long, glabrous to pubescent, eglandular or
occasionally biglandular. Inflorescence 9-17 cm long, simple, with
flowers small, numerous, crowded and all bisexual; peduncle 1-3 cm
long, densely pubescent; rhachis 8-14 cm long, densely pubescent.
Flowers 5-merous, 2.5-4 x 3-4 mm; lower hypanthium 1.5-2.5 mm long,
pubescent to densely so, upper hypanthium campanulate, 1-2 mm long,
pubescent; calyx lobes revolute, 0.8-1.5 mm long, pubescent; stamens 3-
5 mm long; disk pubescent; style 2-4 mm long, densely villous for at
least basal half, sometimes almost so to apex. Infructescence with fruits
often disposed along whole length of rhachis; fruit subglabrous at
maturity, 1-2.5(-3) x 2-4(-5.5) cm, flattened, transversely elliptic in side
view, apex usually emarginate, base truncate or very obtuse, pseudostipe
0.5-3 mm long, wings 2, fairly stiff, equal, 0.8-2.5(-3) cm wide, rounded
to narrowly rounded or rarely pointed laterally, body 0.4-0.8 cm wide,
ridged on one face, flat or depressed on other.

Distribution: Widespread from S Mexico to Bolivia and Bahia in
Brazil, in continental S America much commoner in the west; tall dense
humid primary forest, both inundated and on terra firme, on clays,
limestones and alluvium, on low hills, slopes and ridges and in flood-
plains, also in secondary forest, commonly up to 500 m elev.; the localities
in the Guianas are extreme eastern outliers (GU: 1; SU: 1; FG: 1).

Specimens examined: Guyana: Rupununi Distr., Kanuku Mts., Crabwood Cr., Jansen-Jacobs *et al.* 3407 (LTR, U). Suriname:Nickerie, Fallawatra, Jiménez-Saa 1513 = LBB 14248 (LTR, U). French Guiana: no locality, Grenand 1303 (CAY).

N o t e : Mainly a western species in S America. There is some doubt about the identification of the Guianan species mentioned above, but for the moment they are best placed here; similar examples occur in Pará, Brazil.

7. **Terminalia quintalata** Maguire in Maguire *et al.*, Bull. Torrey Bot. Club 75: 649. 1948. Type: Guyana, Essequibo, Potaro R., below Amatuk portage, Maguire & Fanshawe 23551 (holotype NY, isotypes BM, BR, F, FDG, G, K, MO, P, U, US).

Shrub or tree 0.5-30 m. Leaves strongly coriaceous, obovate or sometimes oblong-obovate or narrowly so, to 4-22 x 2-11.5 cm, apex rounded or retuse or sometimes obtuse, base attenuate-cuneate, glabrous except sometimes sparsely pubescent on midvein below; domatia absent; venation brochidodromous, midvein stout, flattened and scarcely prominent, secondary veins 6-12 pairs, moderately spaced to distant, arising at widely acute to almost right angles, slightly curved to recurved, often forked and joining a well-marked intramarginal vein, scarcely or not prominent, intersecondary veins often present, often almost as evident as secondaries; tertiary veins inconspicuous, randomly reticulate; higher order veins indistinct; areolation imperfect or incomplete, not prominent; petiole 0.3-1.5 cm long, glabrous to sparsely pubescent, eglandular or biglandular at junction with leaf. Inflorescence 7-15(-18) cm long, simple, andromonoecious, sometimes bisexual and sometimes male flowers predominating, the males dispersed among the bisexuals; peduncle 4-7.5 cm long, glabrous to pubescent; rhachis 3-10(-12) cm long, glabrous to pubescent. Flowers (4- to) 5-merous, bisexual ones 5-10.5 x 3.5-5.5 mm, male ones 3-5 x 3.5-5.5 mm; lower hypanthium 2-6.5 mm long in bisexual flowers, pedicel-like and 1-3 mm long in male flowers, glabrous to densely appressed-pubescent, upper hypanthium campanulate to cupuliform, 1.5-3 mm, glabrous to appressed-pubescent outside, always densely pubescent inside; calyx lobes suberect to recurved, 0.8-1.8 mm long, glabrous to appressed-pubescent; stamens 3-6.5 mm long; disk villous to densely so; style 3.5-9.5 mm long, glabrous or very sparsely pubescent near base. Infructescence with numerous fruits, disposed along whole length of rhachis; fruit glabrous to sparsely pubescent, (0.6-)0.8-1.2(-1.5) x 0.5-1(-1.2) cm, actinomorphic, broadly elliptic in side view, apex and base rounded to truncate, wings (4-)5, flexible, equal, 0.2-0.6 cm wide, rounded laterally, body 0.2-0.35 cm wide.

Distribution: Headwaters of the Orinoco, Essequibo and Amazon rivers in Venezuela, Guyana and Brazil; wide range of habitats in a small area, seasonally dry to wet riverine forest, rocky upland scrub, dwarf (4-6 m) forest, cerrado, savanna, at 300-1000 m elev. in Guyana; 13 specimens from the Guianas (GU: 13).

Selected specimens: Guyana: Potaro-Siparuni Region, Kaietur National Park, Kelloff 1058 (K, U, US); Cuyuni-Mazaruni Region, Pakaraima Mts., 0.5 km NW of Imbaimadai settlement, Hoffman 3419 (LTR, U, US); Pakatuk Path, Potaro R., Jenman 7455 (K, U).

Vernacular names: Guyana: fukadi (this name is used for several species of *Terminalia* and *Buchenavia*).

Phenology: Flowering September to June; fruiting October to May.

113. DICHAPETALACEAE

by

GHILLEAN T. PRANCE[6]

Trees, shrubs, lianas. Leaves simple, alternate, entire, pinnately veined; small stipules present but usually caducous. Inflorescences axillary or more frequently attached to petiole or rarely to midrib, corymbose-cymose or subcapitate, or flowers fasciculate. Flowers small, hermaphrodite or less frequently unisexual, actinormorphic to weakly zygomorphic; pedicels often articulated; petals 5, either free, imbricate and almost equal or connate into a tube, lobes equal or markedly unequal, lobes usually bifid at apex and frequently bicucullate or inflexed, often clawed at base; stamens 5, all fertile or only 3 fertile, free or adnate to corolla tube, with filaments or rarely sessile anthers, anthers 2-locular, dehiscing longitudinally; disk of 5 equal or unequal hypogynous glands alternating with stamens or united into a disk; ovary superior, free, 2-3-locular, ovules anatropous, pendulous, paired at top of each locule, styles 2-3, free or more frequently connate nearly to apex, often recurved, stigma capitate or simple. Fruits dry or rarely fleshy drupes, epicarp most frequently pubescent, mesocarp thin, endocarp hard, 1-2(-3)-locular, locules usually with only 1 seed developing; seed pendulous, without endosperm, embryo large, erect.

Distribution: A tropical family of about 240 species in 3 genera, distributed throughout the lowland tropical regions of both hemispheres (but absent from Polynesia and Micronesia), extending into the subtropics in Africa and India; 2 genera and 7 species have been recorded from the Guianas (the third genus, *Stephanopodium* Poepp., does not occur in the Guianas, but in Colombia, Peru, Venezuela and eastern central Brazil).

LITERATURE

Lindeman, J.C. 1986. Dichapetalaceae. In A.L. Stoffers & J.C. Lindeman, Flora of Suriname, additions and corrections 3(2): 548-549.

Prance, G.T. 1971. Dichapetalaceae. In T. Lasser, Flora de Venezuela 3(1): 55-74.

Prance, G.T. 1972. Dichapetalaceae. Flora Neotropica Monograph 10: 1-84.

[6] Royal Botanic Gardens, Kew, Richmond, Surrey, TW9 3AB, U.K.

Prance, G.T. 1980. Dichapetalaceae. In G. Harling & B. Sparre, Flora of Ecuador 12: 1-13.

Prance, G.T. 1998. Dichapetalaceae. In J.A. Steyermark *et al.*, Flora of the Venezuelan Guayana 4: 666-671.

Prance, G.T. 2001. Dichapetalaceae. In R. Bernal & E. Forero, Flora of Colombia 20: 1-61.

Prance, G.T. 2002. Dichapetalaceae. In S.A. Mori *et al.*, Guide to the Vascular Plants of Central French Guiana. Part 2. Mem. New York Bot. Gard. 76(2): 247-249.

Stafleu, F.A. 1951. Dichapetalaceae. In A.A. Pulle, Flora of Suriname 3(2):166-172.

KEY TO THE GENERA

1   Inflorescence with a long distinct peduncle; petals free and regular; stamens free . . . . *1. Dichapetalum*
Inflorescence sessile or almost so; petals connate or only 3, free; stamens adnate to corolla tube 2

2   Corolla with 5 equal obtuse lobes, shorter than tube; fertile stamens 5, anthers sessile on tube (not in the Guianas, see family distribution) . . . . *Stephanopodium*
Corolla zygomorphic, lobes bifid and bicucullate, exceeding tube in length; fertile stamens 3 or 5, anthers on slender filaments . . . . *2. Tapura*

1. **DICHAPETALUM** Thouars, Gen. Nov. Madag. 23. 1806.
Type: D. madagascariense Poir.

Small trees, shrubs or lianas. Leaves petiolate. Inflorescences axillary or adnate to petiole, branched cymose or corymbose panicles with long peduncles. Flowers hermaphrodite, polygamous or dioecious, actinomorphic; bracts small; receptacle usually convex or subplanate; sepals 5, imbricate, free, or connate at base; petals 5, free to base, alternate with sepals, usually bicucullate and 2-lobed at apex, margins of lobes inflexed and sometimes enveloping anthers; stamens 5, equal, free, all fertile in male and hermaphrodite flowers, filaments usually free, rarely connate at base only, anthers broadly oblong, introrse; disk usually consisting of 5 hypogynous glands opposite to petals, glands entire, shallowly lobed, free or united; ovary globose, 2-3-locular with 2 ovules in each locule, styles 1-3, free or connate almost to apex, rudimentary pistil present in male flowers. Fruits dry indehiscent drupes, 1-3-locular, usually with 1 seed in each loculus, epicarp pubescent.

Distribution: Throughout the lowland tropics in Malesia, tropical and southern Africa, and in the New World from Mexico to Peru and Amazonian Brazil, but not in the Caribbean (most diverse and abundant in Africa); 3 of the 20 Neotropical species occur in the Guianas.

Etymology: The genus is named for the petals that are divided often in 2 cucullate lobes.

KEY TO THE SPECIES

1   Leaf blade thickly coriaceous, densely hirsute beneath, venation slightly
    impressed . . . . 2. *D. rugosum*
    Leaf blade coriaceous or chartaceous, glabrous or with few stiff appressed
    hairs only beneath, venation plane 2

2   Leaf blade narrowly oblong to oblong-lanceolate (length breadth ratio
    greater than 2.5); disk lobes small and flattened; style undivided . . . .
    . . . . *3. D. schulzii*
    Leaf blade ovate-elliptic to oblong (length breadth ratio less that 2.2); disk
    lobes large and swollen; style 3-fid . . . . *1. D. pedunculatum*

1. **Dichapetalum pedunculatum** (DC.) Baill., Hist. Pl. 5: 140. 1874. –
    Chailletia pedunculata DC., Nouv. Bull. Sci. Soc. Philom. Paris 2:
    205. 1811. Type: French Guiana, Patris s.n. (holotype G-DC).
    – Fig. 11

    Symphyllanthus glaber Vahl, Kongel. Danske Vidensk. Selsk. Skr. 6: 89.
    1810. – *Dichapetalum glabrum* Prance, Acta Bot. Venez. 3: 304. 1968, non
    Elmer 1908. Type: French Guiana, Richard s.n. (holotype C, herb. Vahl,
    isotypes G, P).

Liana; young branches tomentellous to glabrous. Stipules persistent, linear, to 10 mm long, puberulous to tomentellous, margin entire. Petiole 4-12 mm long, shortly tomentellous at least when young; blade coriaceous, ovate-elliptic to oblong, 5-16 x 2-7.5 cm, acuminate at apex, subcordate to subcuneate and slightly unequal at base, plane not bullate above, glabrous or with a few stiff hairs on venation only beneath; midrib more or less plane above, pubescent to glabrous, 7-9 pairs of secondary veins, more or less plane above, venation plane above. Inflorescence a spreading panicle, usually petiolar, sometimes terminal; rachis and branches tomentellous to glabrescent; peduncle 1-3.5 cm long. Flowers hermaphrodite; pedicel 0.2-1.5 mm long; bracts and bracteoles lanceolate, 0.5-2 mm long, persistent, tomentose; calyx 1.5-2 mm long,

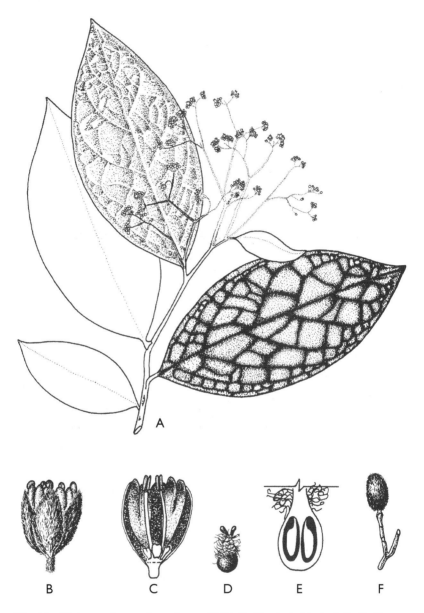

Fig. 11. *Dichapetalum pedunculatum* (DC.) Baill.: A, habit (x 0.5); B, flower (x 8); C, flower section (x 9); D, ovary and style (x 9); E, ovary section (x 16); F, young fruit (x 0.55). Drawing by David Woolcott; adapted with permission from Flora Neotropica Monograph 10.

tomentose on exterior, lobes almost equal; petals of 5 equal lobes, deeply bifid, yellow or white, glabrous; fertile stamens 5, alternating and equalling or shorter than petal lobes; disk of 5 large glands with lobed apices; ovary lanate on exterior, styles lanate at base, glabrous at apex, divided into 2 or 3, or united with a 3-fid apex. Fruit ellipsoid, most frequently 1-locular, but often 2-locular, epicarp short-tomentellous, mesocarp thin, endocarp thin, hard, bony, glabrous within.

Distribution: From Trinidad and northern Venezuela through the Guianas to the western part of Brazilian Amazonia; primary and secondary forests, and river margins, up to 1300 m elev.; many collections studied (GU: 43; SU: 14; FG: 23).

Selected specimens: Guyana: Berbice-Corentyne Region, Corentyne R. above Cow Falls, near Baba Grant Sawmill, McDowell 2338 (K, US); Potaro-Siparuni Region, vicinity of Paramakatoi, Hahn 5661 (K, NY). Suriname: Kalebo Dam near Nickerie, Lindeman & Görts et al. 375 (U, US); Brownsberg, BW 3219 (K, MO, NY, U). French Guiana: Cabassou, 7 Km SW of Cayenne, Hladik 3095 (CAY, NY); Chemin de Vidal, Rémire, Leeuwenberg 11638 (NY, WAG).

Phenology: Flowering around the year.

Vernacular names: Guyana: kahakudiballi rope, massari (Arawak).

Note: De Candolle, in his original description of *Chailletia pedunculata*, did not give any details about this type specimen except to say that it was from Cayenne. Baillon in 'Flora Brasiliensis' cited one of the Paris sheets as the type: Herb. Olyssip s.n., from Pará, Brazil. This sheet came from the herbarium Lusitania in Lisboa, and was presumably collected by A.R. Ferreira in Brazil between 1783 and 1792. It could have been at Paris prior to De Candolle's description of *C. pedunculata*, and was possibly seen by him. However, it disagrees with de Candolle's description in two features, firstly in the locality, Brazil not Cayenne, and secondly in the mature flowers. De Candolle was unable to describe mature flowers of *C. pedunculata*, yet the Herb. Olyssip material has many open flowers. Baillon was certainly incorrect in his designation of the type collection of *D. pedunculatum*. De Candolle in the 'Prodromus' gave a synoptic description of *D. pedunculatum* which included the citation of a single collection by Patris. Thus the type is certainly the Patris collection. The De Candolle herbarium at Geneva has three sheets of *D. pedunculatum* all without field notes or collectors' names. One of these sheets is, however, marked Cayenne. This is almost certainly the type collection, and at least one of the three sheets in the De Candolle herbarium, possibly all three, is the Patris collection.

104

2. **Dichapetalum rugosum** (Vahl) Prance, Acta Bot. Venez. 3: 303.
1968, as 'rugosus'. – *Symphyllanthus rugosus* Vahl, Kongel. Danske
Vidensk. Selsk. Skr. 6: 88. 1810. Type: French Guiana, Herb. Vahl.
s.n. (holotype C).

Cordia scandens Poir. in F. Cuvier, Dict. Sci. Nat. 10: 410. 1818. –
Dichapetalum scandens (Poir.) I. M. Johnst., J. Arnold Arbor. 16: 44. 1935.
Type: French Giuana, Martin s.n. (holotype P).
Chailletia vestita Benth., Hooker's J. Bot. Kew Gard. Misc. 3: 372. 1851.
– *Dichapetalum vestitum* (Benth.) Baill. in Mart., Fl. Bras. 12(1): 371.
1886. Type: Brazil, Pará, Santarém, Spruce 623 (lectotype K, isolectotype
P); Brazil, Pará, Santarém, Spruce s.n. (probably isolectotypes BM,
NY, W).
Dichapetalum vestitum (Benth.) Baill. var. *scandens* Benth. ex Baill. in
Mart., Fl. Bras. 12(1): 372. 1886. Type: Brazil, Amazonas, Manaus, Martius
2794 (lectotype M).
Dichapetalum flavicans Engl., Bot. Jahrb. Syst. 23: 145. 1896. Type:
Guyana, Ri. Schomburgk 1319 (holotype B, lost; photos F, MO).

Liana or shrub; young branches tomentose to tomentellous, becoming
glabrous with age. Stipules lanceolate, 2-4 mm long, subpersistent or
caducous, densely tomentose, margin entire. Petiole 2-35 mm long,
densely tomentose; blade thickly coriaceous, oblong to ovate-elliptic, 6-
32 x 3.5-21 cm, most frequently acute but varying from rounded to
acuminate at apex, subcuneate, rounded or subcordate at base, plane not
bullate, or weakly bullate only above, densely hirsute beneath; midrib
plane and pubescent above, 7-13 pairs of secondary veins, slightly
impressed above. Inflorescence a terminal, axillary or petiolar
corymbose panicle, 1.5-8 cm long, rachis and branches tomentose.
Flowers hermaphrodite; pedicels 0.5-2 mm long; bracts and bracteoles
triangular, 0.5-3 mm long, persistent, tomentose; calyx 3-3.5 mm long,
densely ferrugineous-tomentose on exterior, lobes equal; petals of 5
equal lobes, bifid and weakly cucullate at apex, creamy white, glabrous;
fertile stamens 5, alternating with and equalling petal lobes; disk of 5
separate bifid glands, united at base; ovary lanate on exterior, style lanate
at base, glabrous above, apex shortly 3-fid. Fruit 1-2-locular, epicarp
densely velutinous tomentose, mesocarp thin, endocarp very thin, hard,
bony, glabrous within.

Distribution: The Guianas and Amazonia, and W to the foothills of
the Andes in Peru and Colombia; primary and secondary forests on non-
flooded ground, savannas, and stream margins, up to 500 m elev.; many
collections studied (GU: 3; SU: 5; FG: 6).

Selected specimens: Guyana: Upper Essequibo Region, Rewa R., Clarke 3733 (K, US). Suriname: Commewijne Distr., Perica R., Lindeman 5394 (U); Tafelberg, Maguire 24658 (F, NY, GH). French Guiana: Camopi, Oldeman 3048 (CAY); Maroni, Melinon 233 (P).

Phenology: Collected in flower continuously in Amazonia.

Note: The type of this species is the specimen in the Vahl herbarium at Copenhagen which does not bear a collector's name. It is probably a duplicate of a Richard collection with a *Symphyllanthus* manuscript name which is at Paris and Geneva.

3. **Dichapetalum schulzii** Prance, Bull. Torrey Bot. Club 106: 309. 1979. Type: Suriname, Lower slopes of Frederik Top, 2 km SE of Juliana Top, Irwin *et al.* 54582 (holotype NY, isotypes AAU, U).

Shrub 1.5 m tall; young branches sparsely puberulous. Stipules small, ca. 1 mm long, lanceolate, hirtellous pubescent, margin entire. Petiole 3-7 mm long, terete, sparsely pubescent; blade chartaceous, narrowly oblong to oblong-lanceolate, 7-11 x 2.8-4 cm, glabrous above except on midrib, almost glabrous beneath but with a few scattered stiff appressed hairs, acumen finely pointed slightly curved at apex, acumen 1-2 cm long, rounded and slightly unequal at base; midrib plane and with dense pubescence of stiff appressed hairs above, prominent and with sparse pubescence of stiff appressed hairs beneath, 7-9 pairs of secondary veins, widely spaced, arcuate and anastomosing 4-6 mm from margin, prominulous on both surfaces. Inflorescence a terminal few-flowered panicle, to 2 cm long in bud; rachis and branches brown tomentose; peduncle ca. 1 cm long in young inflorescences. Flowers hermaphrodite, ca. 2.5 mm long; pedicels 1.5 to 2 mm long; bracts and bracteoles small, lanceolate, 1-1.5 mm long, persistent, tomentose; calyx ca. 1 mm long, grey tomentellous on exterior; petals 5, deeply bifid, 2 lobes joined only at extreme base, white, glabrous; fertile stamens 5, alternating and equalling or shorter than petals, enclosed by 1 lobe of each petal in bud, filaments broad, flattened; disk of 5 small glands flattened against filament bases; ovary 2-locular, hirsute on exterior, styles united, with a rounded undivided apex, hirsute. Fruit not seen.

Distribution: Endemic to Suriname, known only from the type collection of a forested ridge at 325 m elev. (SU: 1).

2. **TAPURA** Aubl., Hist. Pl. Guiane 1: 126, t. 48. 1775.
Type: T. guianensis Aubl.

Trees or shrubs. Inflorescences of densely crowded glomerules adnate to petioles or midrib, sessile or subsessile. Flowers small, hermaphrodite or polygamous, weakly zygomorphic; bracts small and often scale-like; sepals 5, imbricate, connate at base, usually unequal rarely equal; petals 5, connate at base to form a long distinct tube, or free almost to base, lobes imbricate, usually with 2 much larger broad lobes which are bifid and bicucullate at apex, and 3 smaller linear-lanceolate entire lobes, rarely with 3 equal or nearly equal bicucullate lobes only; stamens 5, all fertile or more frequently 3 fertile and 2 reduced to staminodes, filaments adnate to inside of corolla tube or to base of corolla in species without a distinct tube, anthers introrse; disk semi-annular or 2-3-partite; ovary free, globose, 2-3-locular with 2 ovules in each loculus, style single, 2-3-lobed at apex or divided for much of length. Fruits dry coriaceous drupes, 1-3-locular with 1 seed in each loculus.

Distribution: Greater and Lesser Antilles, S America from Pacific coastal Colombia through the Guianas and common in Amazonia, also 5 species in Africa; 4 of the 19 Neotropical species occur in the Guianas.

KEY TO THE SPECIES

1 Leaf undersurface hirsutulous. . . . . *1. T. amazonica*
 Leaf undersurface glabrous or with a few stiff appressed hairs only 2

2 Inflorescence with a distinct peduncle of 0.1-1.2 cm long . . . *4. T. singularis*
 Inflorescence sessile on petioles 3

3 Fertile stamens 5, staminodes absent . . . . *2. T. capitulifera*
 Fertile stamens 3, staminodes 2 . . . . *3. T. guianensis*

1. **Tapura amazonica** Poepp. in Poepp. & Endl., Nov. Gen. Sp. 3: 41, t. 246. 1843. Type: Brazil, Amazonas, Tefé, Poeppig 2673 (holotype W, isotype F).

Tapura ciliata Gardn., Icon. Pl. 5: ad t. 466. 1842. – *Tapura amazonica* Poepp. var. *ciliata* (Gardn.) Baill. in Mart., Fl. Bras. 12(1): 375. 1886. Type: Brazil, Goiás, Gardner 3087 (holotype K, isotypes BM, CGE, F, G, GH, NY, OXF, P, W).

Tapura amazonica Poepp. var. *cuspidata* Baill. in Mart., Fl. Bras. 12(1): 375. 1886. Type: not designated.

Tapura amazonica Poepp. var. *dasyphylla* Baill. in Mart., Fl. Bras. 12(1): 375. 1886. Type: not designated.

Tapura amazonica Poepp. var. *sublanceolata* Baill. in Mart., Fl. Bras. 12(1): 375. 1886. Type: not designated.

Tree to 30 m tall, usually much smaller; young branches fulvous-tomentose, becoming glabrous with age. Stipules triangular, 2-4 mm long, pubescent, subpersistent. Petiole 6-16 mm long, tomentose, canaliculate; blade thickly coriaceous, elliptic to obovate-oblong or oblong, 3-25 x 3-9 cm, obtuse to shortly acuminate at apex, acumen 0-10 mm long, rounded to cuneate and often slightly unequal at base, usually plane rarely slightly bullate above, sparsely to densely hirsutulous beneath; midrib impressed above, prominent and pubescent when young beneath, 8-22 pairs of secondary veins, arcuate, anastomosing near margin. Inflorescence a dense glomerule on upper portion of petiole. Flowers hermaphrodite, sessile or pedicels 0.25-2 mm long; bracteoles 0.5-1 mm long, persistent, tomentose; calyx 3.5-5.5 mm long, tomentose on exterior, lobes unequal; corolla yellowish, exceeding calyx lobes, with 2 larger bicucullate lobes and 3 smaller simple lobes, united at base into a very short tube, tube glabrous on exterior, filled by a lanate mass of hair within; fertile stamens 3, alternating with corolla lobes, and inserted at mouth of short corolla tube, 2 staminodes present; ovary 3-locular, pilose on exterior, style with 3-fid apex, pubescent throughout. Fruit oblong-ellipsoid, to 3 cm long, 1-2-locular, epicarp shortly appressed velutinous pubescent, mesocarp 1-4 mm thick, endocarp thin, hard, bony, glabrous within.

Distribution: French Guiana, Amazonia, and the Planalto of Central Brazil; forest on non-flooded ground, up to 350 m elev.; many collections studied (SU: 2 sterile; FG: 15).

Selected specimens: Suriname: Emmaketen, (st.), Daniëls & Jonker 1088 (U); Nassau Mts., (st.), Lanjouw & Lindeman 2409 (U). French Guiana, Cr. Baboune near Sauts, Cremers 7469 (NY, P); Région de Paul Isnard, Mt. Lucifer, Feuillet 309 (CAY, NY).

Phenology: Flowering throughout the year.

Note: In Prance (1972) this species was divided into two varieties. Only var. *amazonica* occurs in the Guianas.

2. **Tapura capitulifera** Baill., Adansonia 11: 112. 1873. Type: Venezuela, Terr. Amazonas, Río Casiquiari, Spruce 3188 (holotype P, isotypes BM, BR, C, CGE, G, GH, GOET, K, LE, NY, OXF, P, RB, W). – Fig. 12

Tree to 15 m tall; young branches tomentellous, soon becoming glabrous. Stipules triangular-lanceolate, ca. 1.5 mm long, pubescent, caducous. Petiole 5-15 mm long, tomentellous when young, becoming less so with age, canaliculate; blade coriaceous, oblong to obovate-lanceolate, 5-12 x 1.8-5 cm, acuminate at apex, acumen 3-10 mm long, subcuneate and slightly unequal at base, glabrous or with a few stiff appressed hairs only beneath, papillose; midrib slightly impressed above, prominent and with a sparse appressed pubescence beneath, 9-15 pairs of secondary veins, arcuate, anastomosing. Inflorescence a dense-flowered glomerule inserted on upper portion of petiole. Flowers hermaphrodite, sessile or subsessile; bracteoles ovate, ca. 0.5 mm long, persistent, pubescent; calyx 2-3 mm long, grey-tomentellous on exterior, lobes unequal; corolla white or yellow, slightly exceeding calyx lobes, with 2 large bicucullate lobes and 3 slightly smaller simple lobes, united at base to form a short tube, tube glabrous on exterior, filled by a dense lanate mass within; fertile stamens 5, alternating with corolla lobes, filaments inserted on corolla tube at base of lobes, lanate at base, staminodes absent; ovary 3-locular, pilose on exterior, style with 3-fid apex. Fruit ellipsoid, 12-18 mm long, usually 1-locular, epicarp short-dense-velutinous-tomentose, mesocarp thin, endocarp thin, hard, bony, glabrous within.

Distribution: Venezuela, and the Guianas; forests, up to 350 m elev. (GU: 7; SU: 18; FG: 10).

Selected specimens: Guyana: Demerara-Berbice Region, Tropenbos Field Station, 20 km S of Mabura Hill, road to Kurupukari, Clarke 751 (K, US); A. Thompson's Farm, S of Timehri, Maas & Westra 3609 (NY, U). Suriname: Sectie O, Stahel 133 (A, NY, U); Brownsberg, Tjon Lim Sang, LBB 16277 (NY). French Guiana: SW of Sinnamary, piste de St. Elie, Monot 52 (CAY); Cr. Baboune, de Granville 4798 (CAY, NY).

Phenology: Flowering December to March.

Vernacular names: Guyana: waiaballi. Suriname: sibalidan.

Note: *Tapura capitulifera* is clearly distinct from *T. guianensis* by the floral structure, but sterile herbarium material is often impossible to distinguish. There are some small, but not consistent, vegetative differences

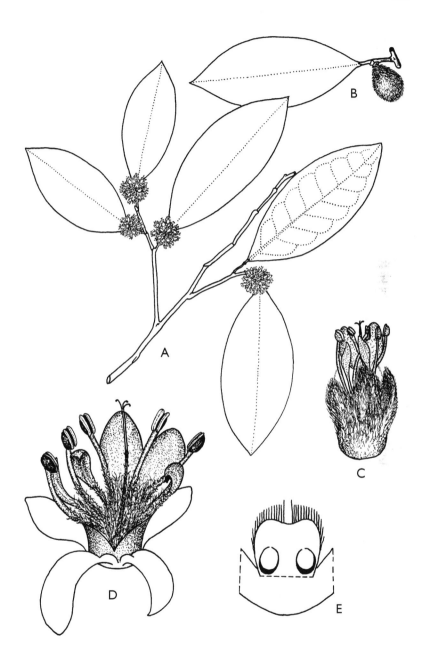

Fig. 12. *Tapura capitulifera* Baill.: A, habit (x 0.65); B, young fruit (x 0.65); C, flower (x 5); D, opened flower (x 6.5); E, ovary section (x 26). Drawing by David Woolcott; adapted with permission from Flora Neotropica Monograph 10.

such as the papillose under surface of the leaf, and the tendency to smaller narrower leaves in *T. capitulifera*. However, in both species there is much variation in leaf shape and size. This is well demonstrated in various Suriname collections where a large number of gatherings have been made from the same numbered tree over several years. Flowering herbarium material is needed to distinguish between *T. guianensis* and *T. capitulifera*. However, Dr. Jan C. Lindeman of Utrecht informs me that these two species are quite easy to distinguish in the field. *T. capitulifera* has a distinct pale dull grey leaf undersurface while both sides of the leaves of *T. guianensis* are green. Also *T. capitulifera* usually develops into a much larger tree.

3. **Tapura guianensis** Aubl., Hist. Pl. Guiane 1: 126, t. 48. 1775. Type: French Guiana, Aublet s.n. (BM, P).

> *Chailletia sessiliflora* DC., Nouv. Bull. Sci. Soc. Philom. Paris 2: 205. 1811. Type: French Guiana, Collector unknown s.n. (G-DC).
> *Tapura cucullata* Benth., Hooker's J. Bot. Kew Gard. Misc. 5: 292. 1853. Type: Brazil, Amazonas, Spruce 2226 (holotype K, isotypes BM, M, P, W).
> *Tapura negrensis* Suess., Repert. Spec. Nov. Regni Veg. 51: 199. 1942. Type: Brazil, Amazonas, Rio Negro, S. Felipe, Luetzelburg 22231 (holotype M).

Tree to 9 m tall or shrub; young branches glabrous or sparsely puberulous, soon becoming glabrous. Stipules lanceolate, to 2 mm long, caducous. Petioles 5-14 mm long, sparsely puberulous to appressed pubescent when young becoming less pubescent with age, rugose, terete to shallowly canaliculate; blade coriaceous, most frequently oblong to ovate-elliptic rarely oblong-lanceolate or lanceolate, 6-23 x 2.1-9 cm, acuminate at apex, acumen 4-18 mm long, rounded to cuneate and unequal at base, glabrous or with a few stiff appressed hairs beneath; midrib impressed above, prominent and glabrous or with a few stiff appressed hairs only beneath, 7-14 pairs of secondary veins, arcuate, anastomosing. Inflorescence a dense sessile glomerule inserted on upper portion of petiole. Flowers hermaphrodite, sessile or on short pedicels; bracteoles 0.5-1 mm long, persistent, pubescent; calyx 3.5-5.5 mm long, tomentellous to sparsely puberulous on exterior, lobes unequal; corolla white or yellow, exserted beyond calyx lobes, consisting of 2 larger bicucullate and 3 smaller simple lobes, united at base to form a long tube, tube glabrescent on exterior, filled by a dense lanate mass within; fertile stamens 3, alternating with corolla lobes, filaments inserted on corolla tube at base of lobes, bases densely pubescent, 2 staminodes present; ovary 3-locular, pilose-tomentose on exterior, style with 3-fid apex, pubescent throughout. Fruit ellipsoid to narrowly oblong, most frequently 1-locular, but often 2-locular, epicarp with short compact velutinous pubescence, mesocarp very thin, endocarp very thin, hard, bony, glabrous within.

Distribution: The Guianas and Amazonia; primary forest on flooded and non-flooded ground, up to 800 m elev.; many collections studied (GU: 92; SU: 60; FG: 86).

Selected specimens: Guyana: Upper Takutu Region, Rupununi area, road from Lethem 25 km past Surama Village, Acevedo Rodr. 3423 (K, US); base of Mt. Makarapan, Makarapan Cr., Maas *et al.* 7507 (K, U). Suriname: Tumac Humac Mts., Talouakem on Litani R., Acevedo Rodr. 5843 (K, US); Brownsberg, Maas *et al.* 2299 (NY, U). French Guiana: R. Mana, Saut Fracas, Ilot du Saut, Cremers 7297 (CAY, NY); Piste de St. Elie, PK 14, Feuillet 197 (CAY).

Phenology: Flowering throughout the year in the Guianas, especially November to April.

Vernacular names: Guyana: lukuchi-danni, muribania,surubundi, waiaballi, waidan. Suriname: boesi kofi, pakira-oedoe, pakira-tiki, wassakao.

Note: For differences with *T. capitulifera*, see note under the latter.

4. **Tapura singularis** Ducke, Trop. Woods 90: 21. 1947. Type: Brazil, Pará, Ducke 1930 (holotype MG, isotypes A, F, GH, IAN, K, NY).

Tree to 25 m tall; young branches shortly tomentellous, glabrescent. Stipules lanceolate, to 4 mm long, persistent. Petioles 3-10 mm long, shortly tomentellous, becoming almost glabrous with age, canaliculated to terete; blade coriaceaous, oblong to oblong-lanceolate, 5-15 x 1.4-4.5 cm, acuminate at apex, acumen 2-12 mm long, subcuneate and equal or slightly unequal at base, glabrous beneath except for a few stiff appressed hairs; midrib slightly impressed above, prominent and stiff appressed pubescent becoming glabrous beneath, 6-8 pairs of secondary veins, arcuate, anastomosing near margin. Inflorescence short axillary cymes, peduncle 0.1-1.2 cm long, tomentellous. Flowers hermaphrodite, pedicels 1.5-3 mm long; bracteoles 0.3-1.5 mm long, triangular, persistent, pubescent; calyx 4-5 mm long, shortly pubescent on exterior, lobes slightly unequal; corolla far exceeding calyx lobes, all lobes united at base to form a long tube, tube sparsely pubescent-glabrescent on exterior, filled with a lanate mass of hair within; fertile stamens 3, alternating with corolla lobes, inserted on corolla tube at base of lobes, 2 staminodes present; ovary 3-locular, tomentose on exterior, style with 3-fid apex, pubescent throughout. Fruit globose to ellipsoid, ca. 2 cm long, with 1 or 2 loculi developing, epicarp with dense compact velutinous pubescence, mesocarp thin, endocarp very thin, hard, bony, glabrous within.

Distribution: Eastern Brazil (Pará, Amapá) and adjacent French Guiana; forests mainly on non-flooded ground, but sometimes in lightly flooded areas; ca. 25 collections studied (FG: 6).

Selected specimens: French Guiana: Station des Nouragues, Sabatier 2530 (CAY, K, P); R. Grande Inini, basin du Maroni, Sabatier *et al.* 3141 (CAY, P, NY).

Phenology: Flowering April to July and fruiting May to January.

167. LIMNOCHARITACEAE

by

ROBERT R. HAYNES[7] & LAURITZ B. HOLM-NIELSEN[8]

Plants herbaceous, perennials, with milky juice, glabrous, growing submersed or with floating leaves, in fresh waters. Roots fibrous, few to many, non-septate, from a stout rhizome or stolon. Stems fleshy, erect, unbranched, internodes without spinulose teeth, tips without turions or tubers. Leaves basal or alternate, petiolate; petioles terete to triangular, mostly 3 or more times length of blade, sheath without auricles, intravaginal scales absent; blades without pellucid markings, margins entire, venation reticulate, with parallel primary veins from base of blade to apex and reticulate secondary veins. Inflorescences scapose, terminal, erect to floating, an involucrate umbel without subtending spathe, involucre of few to several bracts, bracts membranous. Flowers hypogynous, perfect, pedicellate; perianth actinomorphic, of 6 separate segments in 2 series, sepals 3, persistent, mostly erect and enclosing flower and fruit, petals 3, deciduous; stamens 6 to many, separate, anthers 2-locular, basifixed, dehiscing by longitudinal slits; pollen 3-7 aperturate, globose, separate; gynoecium of 3 to many separate carpels, carpels 1-locular, each with numerous anatropous or campylotropous ovules, placentation laminar, styles short or absent, stigma linear. Fruit a follicle, dehiscing adaxially; seeds numerous, glandular pubescent or costate, U-shaped; endosperm helobial in development, absent in mature seed.

D i s t r i b u t i o n : A pantropical family, consisting of 3 genera, of which 2 in the Neotropics, and 8 species, of which 7 in the Neotropics; in the Guianas 1 species.

N o t e : Lemée (Fl. Guyan. Franç. 1: 77. 1955) treats *Limnocharis flava* (L.) Buchenau as occurring in French Guiana, with reference to Aublet (Hist. Pl. Guiane 1: 323. 1775). Aublet only gives a short diagnosis, no plate. Up to the present, no specimen from French Guiana has been found. From its general distribution, however, *L. flava* could be expected in the Guianas.

[7] University of Alabama, Department of Biological Sciences, Box 870345, Tuscaloosa, AL 35487-0345, U.S.A.
[8] University of Aarhus, Rector's office, Nordre Ringgade 1, 8000 Aarhus C, Denmark.

We would like to thank Kirsten Tind and Anni Sloth for preparing the illustrations. This project was supported in part by grant No. 11-4404 from the Danish Natural Science Research Council and by United States National Science Foundation grant INT-8219869.

114

LITERATURE

Haynes, R.R. & L.B. Holm-Nielsen. 1992. The Limnocharitaceae. Flora Neotropica 56: 1-34.

Haynes, R.R. & L.B. Holm-Nielsen. 2001. Limnocharitaceae. In J.A. Steyermark *et al.*, Flora of the Venezuelan Guayana 6: 17-18.

Holm-Nielsen, L.B. & R.R. Haynes. 1986. 192. Limnocharitaceae. In G. Harling & B. Sparre, Flora of Ecuador 26: 25-34.

Jonker, F.P. 1943. Butomaceae. In A. Pulle, Flora of Suriname 1(1): 483-485.

1. **HYDROCLEYS** Rich., Mém. Mus. Hist. Nat. 1: 368. 1815.
Type: H. commersonii Rich.

Plants submersed, with floating leaves. Stems short, erect; stolons often present, terete. Leaves basal, floating or submersed, submersed sessile phyllodia, floating long-petiolate; petioles terete, septate, with a sheathing base; blades orbicular to oblong-lanceolate, margin entire, apex mucronate to obtuse, base rounded to cordate. Inflorescences of few to numerous flowers, terminating a long septate scape, proliferating with leaves and stolons, scapes few to many; peduncles terete, septate; bracts elliptic to lanceolate, delicate, separate, shorter than pedicel subtended. Flowers long-pedicellate, pedicels cylindric, terete; sepals green, coriaceous, erect, elliptic, apex cucullate, sometimes with a midvein; petals yellow to white, delicate, oblong-obovate to orbicular, falling off soon, erect to spreading, longer than or shorter than sepals; stamens 6-many, in 1-several series, outer often sterile, filaments linear or narrowly elliptic, flattened, anthers linear; carpels 5-8, terete, linear-elliptic, basally scarcely cohering, attenuate into style, styles curved inward, papillose at apex. Fruits more or less terete, linear-elliptic, membranous, without dorsal furrows, dehiscing along inner margins; seeds numerous, sparsely glandular pubescent.

Distribution: A genus with its native range restricted to the Neotropics, consisting of 5 species; 1 species in the Guianas.

1. **Hydrocleys nymphoides** (Willd.) Buchenau, Index Crit. Butom. Alism. Juncag. 2. 1868; Abh. Naturwiss. Vereins Bremen 2: 2. 1869. – *Stratiotes nymphoides* Willd., Sp. Pl. 4: 821. 1806. Type: Venezuela, Humboldt & Bonpland s.n. (holotype, B-W, not seen, IDC microfiche 18477). – Fig. 13

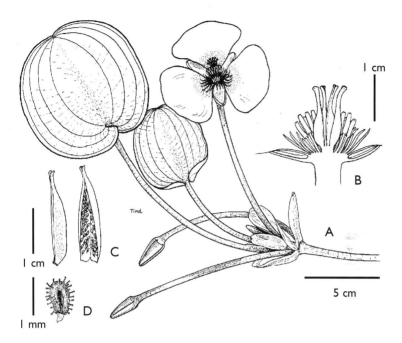

Fig. 13. *Hydrocleys nymphoides* (Willd.) Buchenau: A, habit; B, longitudinal section of flower illustrating two series of stamens and many staminodia; C, fruit, the one on right opened to illustrate laminar placentation; D, seed with glandular pubescence (A-D, from living material cultivated at AAU greenhouse). Reproduced with permission from Haynes & Holm-Nielsen, 1992: 23.

Herbs to 50 cm tall; stolons to 45 cm long. Petiole 1.5-40 cm long, ca. 1 cm wide, sheathing base to 8.5 cm long; blade broadly ovate to orbicular, 1.4-11.9 x 0.9-10.6 cm, apex obtuse to slightly mucronate, base cordate, primary veins 5-9. Inflorescence with 1-6 flowers, proliferating with leaves and stolons; peduncle to 30 cm long, 1.5-6 mm diam; bracts elliptic, 2-4.5 x 0.4-1 cm, apex obtuse; pedicels spreading, 3.5-17.5 cm long, 1.5-6 mm diam. Flowers ca. 6.5 cm wide; sepals 1.3-2.8 x 0.7-1.3 cm, apex obtuse, without midvein; petals pale yellow to white with yellow base, spreading, longer than sepals, 2.3-2.6 x 3.8-4.1 cm; stamens 20-25, in 2 or more series, filaments 5-5.5 mm long, anthers 5.5-6 x ca. 0.5 mm, staminodia numerous; carpels 5-8, ca. 1 cm long. Fruit 1-1.5 x 0.2-0.35 cm, with 0.35-0.55 cm long beak; seeds ca. 1 mm long, sparsely glandular pubescent, glandular trichomes ca. 0.15 mm long, 150-200 m apart, not present on every epidermal cell of seed coat.

Distribution: Widely distributed, mostly outside the Amazon drainage, Guatemala, Colombia, Ecuador, Bolivia, east to the Guianas, eastern and southern Brazil and Argentina; naturalized in the southern United States of America; over 100 collections studied, 22 from the Guianas (GU: 12; SU: 10).

Selected specimens: Guyana: Georgetown, Hitchcock 16619 (GH, NY, S, US); SE of Georgetown, Hekking 1241c (A, NY, U). Suriname: Marowijne, near Hamptencourt-polder, Lanjouw & Lindeman 3194 (NY, U); Saramacca, road from Paramaribo to Carl François, Maguire & Stahel 23590 (BR, F, GH, NY, P, U, US).

Economic use: Cultivated as an aquarium plant or in aquatic rock gardens.

Vernacular name: water-poppy.

168. ALISMATACEAE
by
ROBERT R. HAYNES[9] & LAURITZ B. HOLM-NIELSEN[10]

Plants herbaceous, annual or perennial, with milky sap, glabrous to stellate-pubescent, submersed, floating leaves, or emergent in fresh or brackish waters. Roots fibrous, few to many, septate or aseptate, at base or lower nodes of stem. Stems corm-like, rhizomatous, or stoloniferous, rhizomes occasionally terminated by tubers. Leaves basal, sessile or petiolate; petioles terete to triangular, mostly 2 or more times length of blade, with sheathing base; blades linear, elliptic, ovate to rhomboid, with or without pellucid markings of dots or lines, margin entire or undulate, apex obtuse, acute, or acuminate, base with or without basal lobes, if without basal lobes, then attenuate, if with basal lobes, then truncate, cordate, sagittate, or hastate, venation reticulate, with parallel primary veins from base of blade to apex and reticulate secondary veins. Inflorescences scapose, mostly erect, rarely floating or decumbent, verticillate, forming racemes or verticils branching to become paniculate, rarely umbellate, without a subtending spathe, bracteolate; bracts whorled, linear, delicate to coarse, smooth to papillose, entire, obtuse to acute. Flowers hypogynous, bisexual or unisexual (plants monoecious, rarely dioecious), subsessile to long pedicellate; perianth actinomorphic, of 6 free segments in 2 series, sepals 3, green, persistent, erect and enclosing flower and fruit or spreading to reflexed, petals 3, delicate, larger than sepals, deciduous; stamens absent or of 6 to many, separate, anthers 2-locular, elongate, basifixed or versatile, dehiscing by longitudinal slits; pollen 5-aperturate, globose, separate; gynoecium absent or of 6 to many separate carpels, spirally arranged, carpels 1-locular, each with 1 or rarely 2 anatropous ovules, placentation basal, styles terminal or lateral, persistent, stigmas linear. Achenes compressed or terete, mostly numerous, often winged, longitudinally ribbed or not ribbed, glandular or eglandular; seeds 1-few, U-shaped; endosperm helobial or nuclear in development, absent in mature seeds.

[9] University of Alabama, Department of Biological Sciences, Box 870345, Tuscaloosa, AL 35487-0345, U.S.A.
[10] University of Aarhus, Rector's office, Nordre Ringgade 1, 8000 Aarhus C, Denmark.

We would like to thank Kirsten Tind and Anni Sloth for preparing the illustrations. This project was supported in part by grant No. 11-4404 from the Danish Natural Science Research Council and by United States National Science Foundation grant INT-8219869.

Distribution: A worldwide family of about 75 species in 11 genera; in the Neotropics ca. 40 species in 2 genera; in the Guianas 11 species in 2 genera.

LITERATURE

Agostini, G. 1974. Taxonomic bibliography for the neotropical flora: Alismataceae. Acta Bot. Venez. 9: 269-272.

Bogin, C. 1955. Revision of the genus Sagittaria (Alismataceae). Mem. New York Bot. Gard. 9: 179-233.

Haynes, R.R. & L.B. Holm-Nielsen. 1994. The Alismataceae. Flora Neotropica 64: 1-112.

Haynes, R.R. & L.B. Holm-Nielsen. 1995. Alismataceae. In J.A. Steyermark et al., Flora of the Venezuelan Guayana 2: 377-383.

Holm-Nielsen, L.B. & R.R. Haynes. 1986. 191. Alismataceae. In G. Harling & B. Sparre, Flora of Ecuador 26: 1-24.

Jonker, F.P. 1943. Alismataceae. In A. Pulle, Flora of Suriname 1(1): 472-482

Jonker, F.P. 1968. Alismataceae. In A. Pulle & J. Lanjouw, Flora of Suriname, additions and corrections 1(2): 378-379.

Rataj, K. 1967. Echinodorus intermedius (Martius) Grisebach und verwandte Arten des tropischen Amerika. Mitt. Bot. Staatssaml. München 6: 613-619.

Rataj, K. 1971. The taxonomy of Echinodorus palaefolius (Nees & Mart.) Macbr. (Alismataceae) and related species from Mexico, Central and South America. Preslia 43: 10-16.

Rataj, K. 1972. Revision of the genus Sagittaria. Part II. (The species of West Indies, Central and South America). Annot. Zool. Bot. 78. 61 pp.

Rataj, K. 1975. Revizion [sic] of the genus Echinodorus Rich. Stud. âeskoslov. Akad. Ved 2. 155 pp.

KEY TO THE GENERA

1   Flowers all bisexual; fruits terete, mostly ribbed, with glands between the ribs, not winged, but sometimes (slightly) keeled ...... *1. Echinodorus*
    Flowers, at least the lower ones, unisexual; fruits flattened, often with 1 curved wing and 1 or 2 glands .... *2. Sagittaria*

1. **ECHINODORUS** Rich. ex Engelm. in A. Gray, Manual Bot. 460. 1848.
Type: *E. rostratus* (Nutt.) Engelm. (*Alisma rostratum* Nutt.)

Plants annual or perennial, growing emersed in fresh waters. Roots without septations. Stems rhizomatous. Leaves emersed or submersed; emersed leaves petiolate, petioles mostly triangular, rarely terete, blades linear to broadly ovate, with pellucid markings absent or present as dot or lines; submersed leaves mostly sessile phyllodes, blades mostly linear to rarely ovate, pellucid markings present as dots or lines, or absent. Inflorescences erect, emersed, racemose or paniculate scapes, rarely umbelliform, whorls 1-18; pedicels ascending to recurved. Flowers bisexual, subsessile to pedicellate; sepals herbaceous to coriaceous, without sculpturing, reflexed to spreading or rarely slightly appressed; petals white, larger than sepals; stamens 9 to many, filaments glabrous, anthers versatile or basifixed; gynoecium of many carpels, carpels 1-ovuled, styles terminal or lateral. Achenes terete, often longitudinally ribbed and glandular, rarely with a dorsal keel, separating when mature, beak lateral or terminal, erect, horizontal, or recurved.

Distribution: Native to the Americas, consisting of 27 species distributed from northern U.S.A. to Argentina and Chile; 26 species in the Neotropics; 8 species in the Guianas.

Note: The two specimens of *E. berteroi* (Spreng.) Fassett, listed by Haynes & Holm-Nielsen (1994: 41) as originating from Guyana, have another provenance.

KEY TO THE SPECIES

1   Anthers basifixed; carpels 15-20 2
    Anthers versatile; carpels more than 20 3

2   Emersed leaves with pellucid markings absent; flowers 0.6-0.8 cm wide . .
    . . . . *8. E. tenellus*
    Emersed leaves with pellucid markings present as lines; flowers 0.8-1.7 cm
    wide . . . . *1. E. bolivianus*

3   Fruits eglandular 4
    Fruits glandular 5

4   Pellucid markings of leaf blades present as reticulate network ....
.... *6. E. reticulatus*
Pellucid markings of leaf blades absent .... *5. E. paniculatus*

5   Leaves with basal lobes 6
Leaves without basal lobes 7

6   Inflorescence racemose, decumbant, with 3-4 whorls, vegetatively proliferating; bracts longer than subtending pedicel ..... *3. E. horizontalis*
Inflorescence paniculate or rarely racemose, erect, with 7-14 whorls, not vegetatively proliferating; bracts shorter than subtending pedicel ....
.... *4. E. macrophyllus* subsp. *scaber*

7   Submersed leaves usually present; glands of fruits 5 or more, circular ....
.... *2. E. grisebachii*
Submersed leaves absent; glands of fruits 1-2, rarely to 5, elliptic to linear-elliptic; 8

8   Leaves with pellucid markings as separate lines; rachis usually winged between whorls; fruit beak less than half as long as fruit ....
.... *7a. E. subalatus* subsp. *subalatus*
Leaves without pellucid markings, rarely present as separate lines; rachis triangular between whorls; fruit beak half or more as long as fruit ......
.... *7b. E. subalatus* subsp. *andrieuxii*

1.   **Echinodorus bolivianus** (Rusby) Holm-Niels., Brittonia 31: 276. 1979. – *Alisma bolivianum* Rusby, Mem. New York Bot. Gard. 7: 208. 1927, as 'boliviana'. Type: Bolivia, Reyes, White 1540 (holotype NY, isotypes GH, K, NY, US).   – Fig. 14 (A-E)

Herb, annual, to 45 cm tall, glabrous; rhizome absent; stolon present. Leaves submersed or emersed; emersed leaves pale green to green-brown, petiole terete, 0.4-28.5 x ca. 0.05 cm, basal sheath to 2 cm long, blade linear-elliptic to narrowly elliptic, 1-7.4 x 0.2-1.8 cm, margin entire, apex acute, base acute to attenuate, basal lobes absent, veins 1-3, pellucid markings present as separate lines; submersed leaves pale green, petiole ridged, to 12 x 0.02-0.03 cm, blade linear-elliptic, 3.8-6.5 x 1-1.5 cm, margin entire, apex acute, base acute to attenuate, veins 3, pellucid markings present as separate lines. Inflorescence umbelliform or rarely racemose, erect, with 1-2 whorls, overtopping leaves, not vegetatively proliferating, 5-15 x to 10 cm; whorls with 6-15 flowers; peduncle with 4-5 ridges, not winged, 3.5-31.5 x ca. 0.1 cm; rachis, if present, terete between whorls, not winged; bracts shorter than pedicel subtended, united, connate half of length, delicate, deltoid, 2.8-3.5 x 1-2 mm, apex acute, not dilated basally; pedicels in flower and fruit spreading, 1.1-6.2

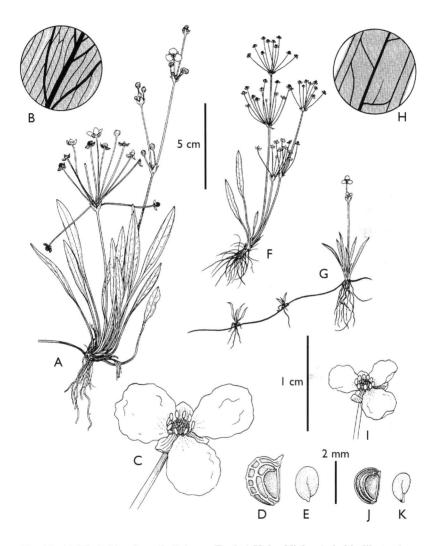

Fig. 14. (A-E). *Echinodorus bolivianus* (Rusby) Holm-Niels.: A, habit, illustrating umbellate inflorescence; B, enlargement of leaf, illustrating pellucid markings as lines; C, flower, illustrating 9 stamens and clawed petals; D, ribbed fruit without glands; E, seed. (F-K). *Echinodorus tenellus* (Mart.) Buchenau: F, habit, illustrating umbellate to racemose inflorescence; G, habit with stolon; H, enlargement of leaf, illustrating absence of pellucid markings; I, flower, illustrating 9 stamens and clawed petals; J, ribbed fruit without beak and without glands; K, seed. (Reproduced with permission from Haynes & Holm-Nielsen 1994).

x ca. 0.05 cm. Flowers 0.8-1.7 cm wide; sepals in flower and fruit spreading, 2.8-3.6 x 2.8-3.6 mm, margin hyaline, with 7 veins, not papillose; petals clawed, 2.5-4.1 x 1.5-3 mm, claw ca. 1 mm long; stamens 9, filaments 1.8-2.1 mm long, anthers basifixed, 0.5-6 mm long, apex obtuse; carpels 15-20. Fruit terete, obovate, 0.8-1.8 x 0.8-1.3 mm, 3-4-ribbed, without keel, eglandular, beak lateral, erect, 0.1-0.3 mm long.

Distribution: Along shores of lakes and small rivers, from S Mexico and Caribbean Islands, S to Argentina, at elevations of 0-2000 m; over 100 collections examined, including 7 from the Guianas (GU: 6; SU: 1).

Specimens examined: Guyana: Ro. Schomburgk add. ser. 163.S (K); 1 mi. E of Onoro mouth, Forest Dept. G397 (NY); Rupununi Savanna, Cook 207a (NY); Rupununi Distr., Wichabai, A.C. Smith 2285 (G, GH, K, MO, NY, S, U); Rupununi Distr., Dadanawa, Jansen-Jacobs *et al.* 2128 (U), 5082 (U, UNA). Suriname: Boven Sipaliwini, kamp XI, Rombouts 345 (U).

Economic use: Sold as aquarium plants under German commercial name "Hahnenfussähnlicher Igelschlauch" and "Hahnenfuss Froschlöffel", and English "Dwarf Amazon Swordplant".

Phenology: Flowers and fruits year round in the tropics.

2. **Echinodorus grisebachii** Small, N. Amer. Fl. 17: 46. 1909. Type: Cuba, Wright 3198 (holotype US, isotypes BM, G, GH, K, NY, S).
– Fig. 15

Herb, short-lived perennial, glabrous, to 100 cm tall; rhizome erect present, to 2 cm; stolon absent. Leaves submersed or emersed; emersed leaves pale green, petiole ridged, 6-25.5 x 0.2-0.35 cm, basal sheath to 5 cm long, blade linear-elliptic to elliptic, 6-22.5 x 1.5-4 cm, margin entire, apex acute to acuminate, base truncate to acute, basal lobes absent, veins 3-7, pellucid markings present as short, separate lines; submersed leaves pale green, petiole triangular, 1.5-7.5 x 0.2-0.3 cm, basal sheath 7 cm long, blade linear-elliptic, 7.5-22.5 x 0.3-1.7 cm, apex acute to acuminate, base truncate to acute, margin entire, veins 1-3, pellucid markings present as short, separate lines. Inflorescence racemose, erect, with 4-12 whorls, overtopping leaves, often vegetatively proliferating, 7-40 x 0.4-1.5 cm; whorls with ca. 6 flowers; peduncle slightly ridged, slightly winged, 5-20 x 0.05-0.2 cm; rachis between whorls terete, not winged; bracts longer to shorter than pedicel subtended, separate, delicate, elliptic, to 5 x to 2 mm, membranous,

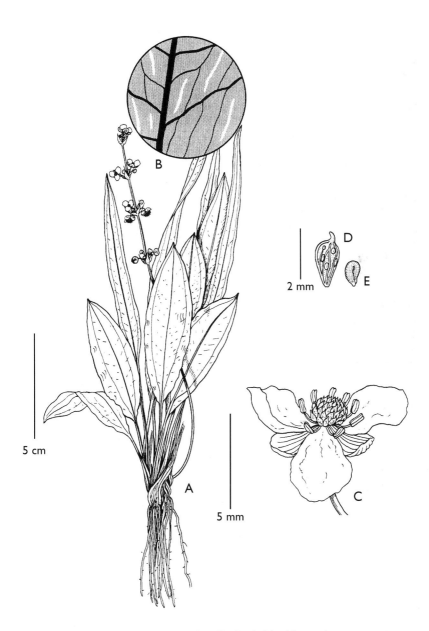

Fig. 15. *Echinodorus grisebachii* Small: A, habit, illustrating emersed and submersed leaves and racemose inflorescence; B, enlargement of leaf, illustrating pellucid markings as separate lines; C, flower, illustrating 8 stamens and petals without claws; D, ribbed fruit with several glands; E, seed. (Reproduced with permission from Haynes & Holm-Nielsen 1994).

124

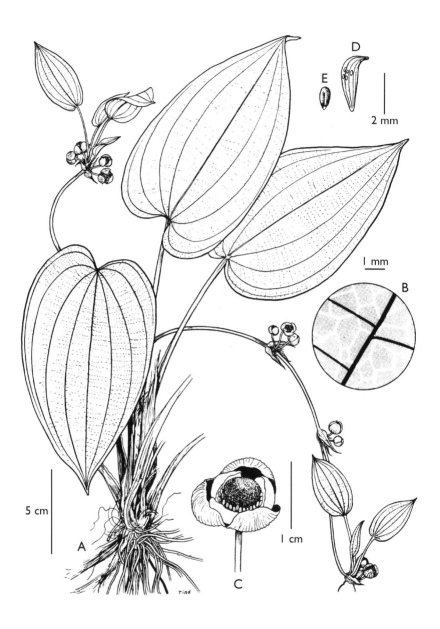

Fig. 16. *Echinodorus horizontalis* Rataj: A, habit; B, enlargement of leaf, illustrating pellucid markings as network; C, flower, showing petals shorter than sepals; D, ribbed fruit with glands in 2 series; E, seed. (Reproduced with permission from Holm-Nielsen & Haynes 1986).

apex acute, not dilated basally; pedicels in flower spreading, in fruit recurved, 0.4-1 x ca. 0.04 cm. Flowers to 1 cm wide; sepals in flower reflexed, 2.5-3 mm long, with ca. 9 veins, not papillose; petals not clawed, ca. 5 x ca. 3 mm; stamens 8-12, filaments ca. 1.5 mm long, anthers versatile, 0.5-1 mm long, apex obtuse; carpels 30 or more. Fruit terete, clavate, 1.8-2.3 x 0.8-1 mm, 3-4-ribbed, without keel, glands 5 or more, circular, between ribs, beak terminal, erect, to 0.5 mm long.

Distribution: In wet ditches and marshes and along shores of lakes and small rivers, from Cuba and Honduras, S to Colombia, Peru, Bolivia, Brazil, and NE to Venezuela and the Guianas, at 0-500 m elev.; 41 collections examined, including 13 from the Guianas (GU: 1; SU: 5; FG: 5).

Selected specimens: Guyana: Rupununi Distr., Dadanawa, Jansen-Jacobs *et al.* 5084 (U, UNA). Suriname: Lawa R., Geijskes 191 (A, NY, U); Litani R., Rombouts 918 (K, NY, U). French Guiana: Antécume Pata, Veth *et al.* 28, 202 (U); Cayenne, Mothery 65 (K); near St. Georges, de Granville 1045 (CAY, MO, NY); sur îlet, pekuatã, Oyapock, Prévost & Grenand 1039 (UNA).

Economic use: Sold under German commercial name "Smalblättrige Amazonas Schwertpflanze" ("Small Leaved Amazon Swordplant."); also cultivated under names "Amazonian Swordplant", or 'E. rangeri', 'E. amphibius', 'E. gracilis', 'E. parviflorus', 'E. bleheri' and 'E. parviflorus tropica'.

Phenology: Flowering and fruiting year round.

Note: See note under *E. subalatus* subsp. *subalatus*.

3. **Echinodorus horizontalis** Rataj, Folia Geobot. Phytotax. 4: 335. 1969. Type: Peru, Loreto, Gamitanacocha, R. Mazán, Schunke 279 (holotype US, isotypes GH, F, NY). – Fig. 16

Herb, perennial, to 150 cm tall, glabrous; rhizome short, erect; stolon present. Leaves emersed, olive-green to greenish brown, petiole terete, 2-4 times as long as blade, 50-60 x 0.5 cm, basal sheath to 10 cm long, blade ovate, 10-25 x 10-15 cm, margin entire, apex acuminate, base cordate, lobes rounded, veins 7-11, pellucid markings present, forming a network independent of veins. Inflorescence racemose, decumbant, with 3-4 whorls, overtopping leaves, vegetatively proliferating, 100 x 3-4 cm; whorls with 3-6 flowers; peduncle terete, 50 x 0.01-0.04 cm; rachis between whorls terete; bracts longer than pedicel subtended, coarse inward, membranaceous along margin, ovate, 30 x 0.3-0.6 mm, apex long-acuminate; pedicels in flower spreading, in fruit reflexed, cylindrical, 2 x 0.05-0.1 cm. Flowers ca. 1.5 cm wide; sepals erect, appressed to mature fruiting aggregate, 10 x 8 mm, 22-28-veined, not papillose; petals not clawed, 5-6 x ca. 4 mm; stamens 24-30, filaments ca. 1.5 mm long, anthers versatile, ca. 1.5 mm long, apex obtuse; carpels numerous. Fruit terete, clavate, 2.5-3 x 0.2-0.8 mm, 3-ribbed, without keel, glands 6-9, circular, irregularly arranged, often 4-6 in upper half and 1-4 in lower half of fruit, beak terminal, recurved, to 0.8 mm long.

Distribution: Colombia, Venezuela, Guyana, S to Peru and Brazil, at 30-1300 m elev.; ca. 25 collections examined, including 1 from Guyana (GU: 1).

Specimen examined: Guyana, NW-Distr., Barama R., Kariako, Awaramu, van Andel *et al.* 1107 (U, UNA).

Economic use: Sold under German commercial name "Horizontale Amazonpflanze" ("Horizontal Amazon plant").

Phenology: Flowering and fruiting year round.

4. **Echinodorus macrophyllus** (Kunth) Micheli in A. DC. & C. DC., Monogr. Phan. 3: 50. 1881. – *Alisma macrophyllum* Kunth, Enum. Pl. 3: 151. 1841. Type: Brazil, Rio de Janeiro, Porto d'Estrella, Sellow s.n. (holotype BM, photo MO).   – Fig. 17 (A-C)

In the Guianas only: subsp. **scaber** (Rataj) R.R. Haynes & Holm-Niels., Brittonia 38: 331. 1986. – *Echinodorus scaber* Rataj, Folia Geobot. Phytotax. 4: 438. 1969. Type: Guyana, Canje R., Jenman 4310 (holotype K).   – Fig. 17 (D-G)

Herb, perennial, to 200 cm tall, glabrous or stellate-pubescent; rhizome present; stolon absent. Leaves emersed, green-brown, petiole ridged, stellate-pubescent to scabrous, 17-41 x 0.3-0.5 cm, basal sheath to 15 cm long, blade ovate to ovate-elliptic, 8.5-41 x 4.9-26 cm, margin entire, apex acute to round-ovate to retuse, base cordate to truncate, basal lobes present, veins 7-11, pellucid markings absent. Inflorescence paniculate or rarely racemose, erect, with 7-14 whorls, overtopping leaves, not vegetatively proliferating; whorls with 8-20 flowers; peduncle ridged, not winged, 28.5-50 x 0.3-1 cm; rachis between whorls triangular, not winged, stellate-pubescent to scabrous; bracts shorter than pedicel subtended, coarse, elliptic, 3-25 x 2.5-4 mm, apex acute, not dilated

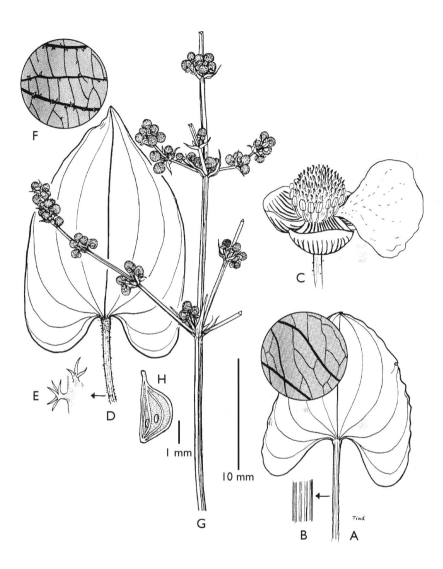

Fig. 17. (A-C). *Echinodorus macrophyllus* (Kunth) Micheli subsp. *macrophyllus*: A, leaf, and enlargement of leaf, illustrating pellucid markings absent; B, enlargement of petiole, indicating glabrous condition; C, flower. (D-G). *Echinodorus macrophyllus* (Kunth) Micheli subsp. *scaber* (Rataj) R.R. Haynes & Holm-Niels.: D, leaf; E, enlargement of petiole, indicating stellate pubescence; F, enlargement of leaf, illustrating pellucid markings absent; G, branched inflorescence; H, ribbed fruit with 2 glands. (Reproduced with permission from Haynes & Holm-Nielsen 1994).

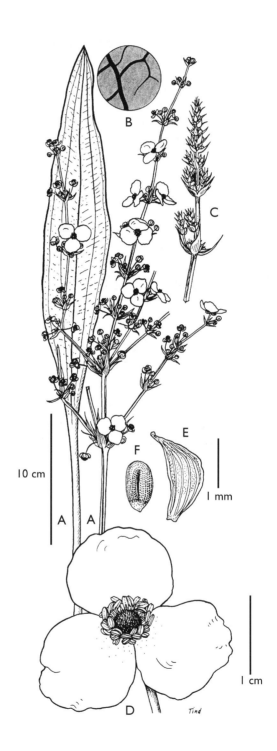

basally; pedicels in flower and fruit spreading, cylindrical to triangular, 0.4-3.5 x 0.03-0.08 cm. Flower 1.5-2.5 cm wide; sepals in flower and fruit spreading, 4.1-4.5 x 3.5-5.8 mm, with 19-25 veins, not papillose; petals not clawed, 6.4-10.5 x 4.5-7.4 mm; stamens 18-21, filaments 0.5-1.5 mm long, anthers versatile, 0.7-1.5 mm long, apex obtuse; carpels 30 or more. Fruit flattened, obovate, 2.3-2.8 x 1-1.5 mm, 2-4-ribbed, slightly 1-keeled, glands 1-2(-5), linear-elliptic, between ribs, beak lateral, erect, 0.5-1 mm long.

Distribution: In marshes and along margins of lakes and ponds from S Nicaragua, Colombia, Venezuela, Guyana and Suriname, S to Bolivia, Brazil and Paraguay, at 50-300 m elev.; 40 collections were examined, 9 from the Guianas (GU: 7; SU: 1; FG: 1).

Specimens examined: Guyana: the type; Berbice, Jenman 5162 (NY), 5272 (NY); near New Amsterdam, Jenman 5080 (NY); Moka Moka Cr., near Kanuku Mts., Goodland 812 (US); Rupununi Distr., Aishalton, Stoffers & Görts et al. 425 (U); Rupununi Distr., Jerome's Place, Jansen-Jacobs et al. 5037 (U, UNA); Upper Takutu Distr., Rupununi R., S of Wichabai, Horn & Wiersema 10103 (NBYC, UNA). Suriname: S of Clara polder, Lanjouw & Lindeman 3144 (NY, U). French Guiana: Richard s.n. (P).

Phenology: Flowering and fruiting the year round.

Notes: The specimen of E. macrophyllus subsp. macrophyllus, listed by Haynes & Holm-Nielsen (1994: 57) as originating from Guyana, is from Brazil.
Jonker (1968: 378) treats E. grandiflorus (Cham. & Schltdl.) Micheli var. floribundus (Seub.) Micheli (= E. grandiflorus subsp. aureus (Fassett) R.R. Haynes & Holm-Niels.) as occurring in Suriname. The only specimen he mentions, however, belongs to E. macrophyllus subsp. scaber.

5. **Echinodorus paniculatus** Micheli in A. DC. & C. DC., Monogr. Phan. 3: 51. 1881. Type: Guyana, Ro. Schomburgk add. ser. 220.S (lectotype K) (designated by Rataj, Bull. Jard. Bot. Natl. Belg. 38: 405. 1968). – Fig. 18

Fig. 18. *Echinodorus paniculatus* Micheli: A, leaf and paniculate inflorescence; B, enlargement of leaf, illustrating pellucid markings absent; C, top of inflorescence; D, flower; E, ribbed fruit without glands; F, seed. (Reproduced with permission from Haynes & Holm-Nielsen 1994).

Herb, annual or perennial, glabrous, to 350 cm tall; rhizome present, to 3 x to 4 cm; stolon absent. Leaves emersed, green to green-brown, petiole triangular, 17.5-55 x 0.5-1.1 cm, basal sheath to 23 cm long, blade narrowly elliptic to elliptic-ovate, 12.5-27 x 1.4-15 cm, margin entire, apex acute, base truncate to attenuate, basal lobes absent, veins 5-7, pellucid markings absent. Inflorescence racemose or paniculate, erect or rarely decumbent, with 4-8 whorls, overtopping leaves, often vegetatively proliferating, to 40 x to 30 cm; whorls with 6-40 flowers; peduncle triangular, not winged, 40-100 x 0.03-0.1 cm; rachis between whorls triangular, not winged; bracts shorter than pedicel subtended, coarse, linear, 15-31 x 0.2-0.5 mm, apex acuminate to subulate, not dilated basally; pedicels in flower and fruit spreading, cylindrical, 0.5-6.5 x 0.05-0.1 cm. Flowers ca. 1.6 cm wide; sepals in flower and fruit spreading, 4.5-6.5 x 1.8-3.2 mm, with ca. 20 veins, mostly not papillose; petals not clawed, ca. 7 x ca. 6 mm; stamens 20 or more, filaments to 1.5 mm long, anthers versatile, to 2.5 mm long, apex obtuse; carpels 30 or more. Fruit terete, obelliptic, 1.5-3 x 0.4-0.8 mm, 4-ribbed, 1-keeled, eglandular, beak terminal, erect, to 0.7 mm long.

Distribution: In marshes and along shores of lakes and streams from Mexico, Nicaragua, and northern S America S to Paraguay and Argentina, from sea level to 1500 m elev.; more than 100 collections examined, 2 from Guyana (GU: 2).

Specimens examined: Guyana: the type; Kanuku Mts., Rupununi R., Bush Mouth near Witaru Falls, Jansen-Jacobs *et al.* 71 (U).

Economic use: Sold as aquarium plants under the German commercial name "Grosse Amazonas Schwertpflanze" ("Large Amazon Swordplant").

Phenology: Flowering and fruiting year round.

6. **Echinodorus reticulatus** R.R. Haynes & Holm-Niels., Brittonia 38: 327. 1986. Type: Suriname, Sipaliwini savanna area on Brazilian frontier, in great "Maurisie" forest, 1.5 km N of "4-Gebroeders" Mts., Oldenburger, Norde & Schulz 292 (holotype NY, isotypes NY, U). – Fig. 19

Herb, perennial, glabrous, to 125 cm tall; rhizome present, stout, short; stolon absent. Leaves emersed, green-brown, petiole terete, to 37 x ca. 0.7 cm, basal sheath to 150 cm long, blade elliptic-ovate, 30-54 x 8.2-12.5 cm, margin entire, apex acute, base attenuate, basal lobes absent, veins 3-5, pellucid markings present as reticulate network. Inflorescence racemose, erect, with ca. 7 whorls, overtopping leaves, not vegetatively

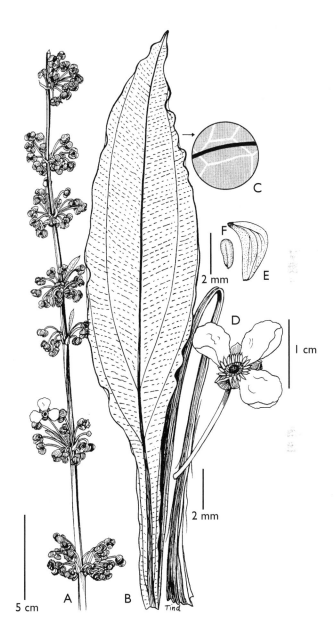

Fig. 19. *Echinodorus reticulatus* R.R. Haynes & Holm-Niels.: A, racemose inflorescence; B, emersed leaf; C, enlargement of leaf, illustrating reticulate pellucid markings; D, flower with acuminate stamens; E, ribbed fruit without glands; F, seed. (Reproduced with permission from Haynes & Holm-Nielsen 1994).

proliferating, to 53 x to 7 cm; whorls with 10-18 flowers; peduncle triangular, not winged, length unknown, ca. 6 mm wide; rachis between whorls triangular, not winged; bracts shorter than pedicel subtended, coarse, elliptic, to 15 x to 0.6 mm, apex long acuminate, not dilated basally; pedicels in flower spreading, in fruit recurved, cylindrical, to 4.5 x to 0.25 cm. Flowers ca. 0.25 cm wide; sepals in flower spreading, in fruit appressed, to 15 x ca. 8 mm, with to 20 veins, not papillose; petals not clawed, spreading, ca. 10.5 x ca. 8 mm; stamens ca. 25, filaments to 2 mm long, anthers versatile, to 1.5 mm long, apex acuminate; carpels 30 or more. Fruit flattened, clavate, 3-3.5 x 1.2-1.5 mm, 4-5-ribbed, slightly 1-keeled, eglandular, beak lateral, horizontal, 1-1.5 mm long.

Distribution: Suriname: only known from the Sipaliwini area in Suriname (SU: 2).

Specimens examined: Suriname: the type; Boven Sipaliwini, kamp XVII, Rombouts 447 (U).

Phenology: Flowers and fruits in February and October.

7. **Echinodorus subalatus** (Mart.) Griseb., Cat. Pl. Cub. 218. 1866. –
    Alisma subalatum Mart. in Schult. & Schult. f., Syst. Veg. 7: 1609.
    1830. Type: Brazil, Martius 150 (lectotype M) (designated by Rataj,
    Preslia 43: 15. 1971).

Herb, short-lived annual, glabrous, to 100 cm tall; rhizome present, to 4 x to 3 cm; stolon absent. Leaves emersed, pale green to green-brown, petiole terete, 6.2-73.4 x 0.1-1.5 cm, basal sheath to 12 cm long, blade linear-elliptic to ovate, 8.5-48 x 1.5-18 cm, margin entire, apex acute, base attenuate to truncate, basal lobes absent, veins 7-11, pellucid markings present as separate lines. Inflorescence racemose or paniculate, erect, with 7-15 whorls, overtopping leaves, not vegetatively proliferating, to 30 x to 2.8 cm; whorls with 3-11 flowers; peduncle semi-terete, not winged, 15-73.5 x 0.2-0.7 cm; rachis between whorls triangular, winged or wingless; bracts longer than pedicel subtended, separate, coarse, subulate, 5-44 x to 0.7 mm, delicate, apex long attenuate, not dilated basally; pedicels in flower and fruit spreading to ascending, cylindrical, 0.5-1.5 x 0.04-0.1 cm. Flowers 1-2 cm wide; sepals in flower and fruit reflexed to spreading, 2.5-4 x 1.8-4 mm, with 11-15 veins, not papillose; petals clawed, 7-10 x 5-9 mm, claw to 0.5 mm long; stamens 15-20, filaments ca. 1.5 mm long, anthers versatile, 1.5-2.5 mm long, apex obtuse; carpels 30 or more. Fruit terete, obelliptic, 1.2-2.1 x 0.6-1 mm, 5-ribbed, without keel, glands 1-2, elliptic, between ribs, beak terminal, erect, 0.7-1 mm long.

7a. **E. subalatus** (Mart.) Griseb. subsp. **subalatus**

Alisma intermedium Mart. in Schult. & Schult. f., Syst. Veg. 7: 1609. 1830.
– *Echinodorus intermedius* (Mart.) Griseb., Cat. Pl. Cub. 218. 1866. Type:
Brazil, Martius s.n. (holotype M).

Petiole with a sheath to 12 cm long; blade with 7-11 veins, pellucid
markings present as separate lines. Inflorescence usually with winged
rachis. Fruit with beak less than half as long as fruit.

Distribution: Marshes and shores of lakes from Venezuela to
Guyana and S to Bolivia, Brazil and Paraguay, from sea level to 500 m
elev.; more than 60 collections examined, 10 from Guyana (GU: 10).

Specimens examined: Guyana: Ro. Schomburgk ser. I, 563 (BM,
G, K); Kanuku Mts., Rupununi R., Bush Mouth near Witaru Falls,
Jansen-Jacobs *et al.* 70 (U); Wasaputa, Cal. Essequibo R., Jenman 1079
(K); Upper Javentua, Appun 2164 (K); Corantyne R., Im Thurn s.n. (K);
Rupununi, Dadanawa, Cook 21 (K, M, U); Takutu R., Lethem, Collector
unknown 388 (K); Rupununi, Chaakoitou, near Mountain Point, Maas &
Westra 4072 (F, NY, MO, U); Rupununi Distr., 15 km S of Aishalton,
Stoffers & Görts et al., 535 (U); UpperTakutu-Upper Essequibo Distr.,
NW of Karanambo, Horn & Wiersema 11072 (NBYC, UNA).

Economic use: Sold as aquarium plants under German commercial
name "Langgriffliger Froschlöffel" ("Longstyled Toadspoon"); also sold
under name E. 'inpai'.

Phenology: Flowering and fruiting year round.

Note: Jonker (1943 : 476) treats *E. intermedius* as occurring in
Suriname. The only specimen that he mentions, however, belongs to *E.
grisebachii*.

7b. **E. subalatus** (Mart.) Griseb. subsp. **andrieuxii** (Hook. & Arn.) R.R.
Haynes & Holm-Niels., Brittonia 38: 327. 1986. – *Alisma andrieuxii*
Hook. & Arn., Bot. Beechey Voy. 311. 1838. – *Echinodorus
andrieuxii* (Hook. & Arn.) Small, N. Amer. Fl. 17: 46. 1909. Type:
Mexico, Andrieux 91 (holotype K). – Fig. 20

Petiole with a sheath to 7 cm long; blade with 7-9 veins, pellucid
markings absent or rarely present as separate lines. Inflorescence with
wingless rachis. Fruit with beak half or more as long as fruit.

134

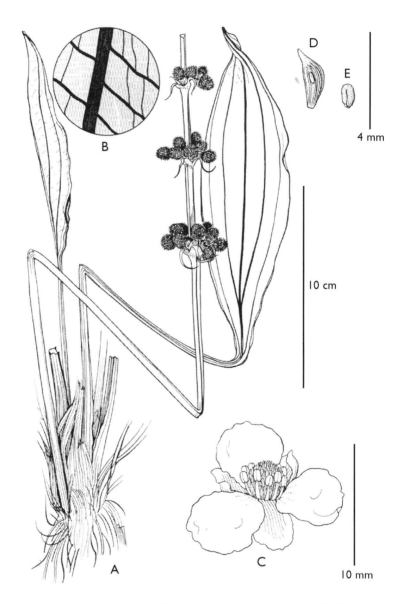

Fig. 20. *Echinodorus subalatus* (Mart.) Griseb. subsp. *andrieuxii* (Hook. & Arn.) R.R. Haynes & Holm-Niels.: A, habit, showing erect inflorescence with triangular rachis; B, enlargement of leaf, illustrating pellucid markings as separate lines; C, flower; D, ribbed fruit with solitary facial gland; E, seed. (Reproduced with permission from Haynes & Holm-Nielsen 1994).

Distribution: Marshes and shores of lakes from S Mexico through C America, Venezuela, Guyana to eastern Brazil, at elevations of sea level to 700 m; more than 100 collections examined, 1 from Guyana (GU: 1).

Specimen examined: Guyana: locality unknown, Appun 2164 (K).

Phenology: Flowering and fruiting year round.

Note: In the literature, there has been much confusion on the type (Haynes & Holm-Nielsen 1994: 37). Application of Art. 37.3 of the ICBN (McNeill *et al.*, Regnum Veg. 146), however, leaves open only one possibility: Hooker & Arnott only mentioned Andrieux 91, this clearly was the specimen on which they based their description. In addition, these authors state that Andrieux had found this species in Oaxaca as well; a comment indicating a second locality, without mention of a second specimen.

8. **Echinodorus tenellus** (Mart.) Buchenau, Index Crit. Butom. Alism. Juncag. 21. 1868; Abh. Naturwiss. Vereins Bremen 2: 21. 1869. – *Alisma tenellum* Mart. in Schult. & Schult.f., Syst. Veg. 7: 1600. 1830. Type: Brazil, Buritihaes, Contendas, Martius s.n. (lectotype M) (designated by Rataj 1975: 13).   – Fig. 14 (F-K)

Herb, annual, glabrous, to 25 cm tall; rhizome absent; stolon present. Leaves submersed, or emersed; emersed leaves pale green to green-brown, petiole terete, 1.2-9.5 x 0.04-0.08 cm, basal sheath 1 cm long, blade linear, 1-7.4 x 0.2-0.5 cm, margin entire, apex acute, base attenuate, basal lobes absent, veins 3-5, pellucid markings absent; submersed leaves pale green, sessile, blade linear, to 5 x ca. 0.2 cm, margin entire, apex acute, base tapering, veins 1-3, pellucid markings absent. Inflorescence umbelliform or rarely racemose, erect, with 1 or 2 whorls, overtopping leaves, not vegetatively proliferating, to 6 x to 8 cm; whorls with 4-6 flowers; peduncle terete, not winged, 12-40 x 0.1-0.2 mm; rachis, if present, between whorls terete, not winged; bracts shorter than pedicel subtended, united, connate half of length, coarse, deltoid, 2.8-4.9 x 1-2.1 mm, delicate, apex acute, not dilated basally; pedicels in flower and fruit spreading, cylindrical, 0.5-3 x 0.03-0.05 cm. Flowers 0.6-0.8 cm wide; sepals in flower and fruit slightly appressed, 2.5-2.9 mm, with 3-5 veins, not papillose; petals clawed, 2.5-4.1 x 1.5-3 mm, claw ca. 1 mm long; stamens 9, filaments ca. 0.5 mm long, anthers basifixed, ca. 1 mm long, apex obtuse; carpels 15-20. Fruit flattened, obovate, 0.8-1.5 x 0.8-1 mm, 0-3-ribbed, without keel, eglandular, beak lateral, erect, 0.1-0.2 mm long.

136

Distribution: Along shores of lakes and small streams from
northeastern U.S.A. to S Brazil, at elevations from 0-1500 m; over 70
collections examined, including 4 from Guyana (GU: 4).

Specimens examined: Guyana: Rupununi, Farne Pan Lake near
Pirara, Graham 389 (K); savanna near Dadanawa Ranch, Görts-van Rijn
et al. 242 (U, UNA); Dadanawa, Jansen-Jacobs et al. 3214 (U); Upper
Takutu-Upper Essequibo, N of Karanambo, Horn & Wiersema 11083
(NBYC, UNA).

Economic use: Sold as an aquarium plant under German
commercial names of "Zartblättriger Froschlöffel" and "Kleinblättriger
Froschlöffel", and English names "Pigmy chain-sword" and "Dwarf
Amazonian Swordplant".

Phenology: Flowers and fruits throughout year in tropics.

2. **SAGITTARIA** L., Sp. Pl. 993. 1753.
   Type: S. sagittifolia L.

   Lophotocarpus T. Durand, Index Gen. Phan. X. 1888.
   Type: L. guayanensis (Kunth) J.G. Sm. (Sagittaria guayanensis Kunth)

Plants monoecious, perennial or rarely annual, glabrous to sparsely
pubescent, growing submersed, floating-leaved, or emersed in fresh or
brackish waters. Roots septate. Stems often with rhizomes; rhizomes
occasionally terminated by tubers, tubers brown, smooth. Leaves
submersed, floating, or emersed, sessile or petiolate, petioles terete to
triangular, blades present or absent, linear to rhombiform to sagittate,
without pellucid markings. Inflorescences erect, racemose or paniculate
scapes, emersed or floating, rarely submersed, whorls 1-17, each whorl
with 2-3 flowers; bracts coarse or delicate, smooth to papillose, obtuse to
acute; pedicels elongating after anthesis, ascending to recurved;
carpellate flowers in lower part of inflorescences, staminate flowers
above. Flowers unisexual, rarely lower ones with a ring of sterile
stamens, pedicellate; sepals herbaceous to coriaceous, often sculptured,
reflexed in staminate flowers, reflexed to appressed in carpellate flowers;
petals white or rarely with a pink to purple spot or tinge; stamens 7-
many, filaments linear to dilated, glabrous to pubescent, anthers
basifixed, linear to orbicular; gynoecium of many carpels, carpels 1-
ovuled, styles terminal. Achenes compressed, with a conspicuous dorsal
wing, not ribbed, glandular.

Distribution: A predominently Western Hemisphere genus of ca. 20 species, from Canada S to Argentina and Chile; 3 or 4 species also occur in Europe and Asia; 3 species in the Guianas.

Note: Bogin (1955) combined *Sagittaria* and *Lophotocarpus*, recognizing 2 subgenera:
subgenus *Sagittaria*: ascending to reflexed sepals and mostly spreading to ascending pedicels in fruit, and all flowers unisexual.
subgenus *Lophotocarpus*: appressed sepals and recurved pedicels in fruit, and upper flowers bisexual.

KEY TO THE SPECIES

1 Pedicels in fruit spreading, not thickened, 0.3-0.4 mm wide; sepals reflexed or spreading from fruiting aggregate . . . . *2. S. lancifolia*
Pedicels in fruit reflexed or spreading, thickened, 1.2-5.5 mm wide; sepals appressed or erect around fruiting aggregate 2

2 Leaves and inflorescences submersed or floating; floating leaves with cordate, sagittate, or hastate base, basal lobes present . . . . *1. S. guayanensis*
Leaves and inflorescences emersed or submersed, not floating; emersed leaves with acute to subcordate base, basal lobes absent . . . . *3. S. rhombifolia*

1. **Sagittaria guayanensis** Kunth in Humb., Bonpl. & Kunth, Nov. Gen. Sp. ed. qu. 1: 250. 1816. – *Echinodorus guayanensis* (Kunth) Griseb., Fl. Brit. W. I. 505. 1862, as 'guianensis'. – *Lophiocarpus guayanensis* (Kunth) Micheli in A. DC. & C. DC., Monogr. Phan. 3: 62. 1881, as 'guyanensis'. – *Lophotocarpus guayanensis* (Kunth) J.G. Sm., N. Amer. Sagittaria 35. 1894, as 'guyanensis'. Type: Venezuela, Bolivar, Humboldt & Bonpland s.n. (holotype P(lost), isotype MO); Suriname, Hostmann 870 (epitype TCD, with duplicates at LE, W).

In the Guianas only: subsp. **guayanensis**   – Fig. 21

Sagittaria echinocarpa Mart., Ausw. Merkw. Pfl. 6. 1829. – *Alisma echinocarpum* (Mart.) Seub. in Mart., Fl. Bras. 3(1): 105. 1847. – *Lophotocarpus guayanensis* (Kunth) J.G. Sm. var. *echinocarpa* (Mart.) Buchenau in Engl., Pflanzenreich IV. 15 (Heft 16): 36. 1903. Type: not designated.

138

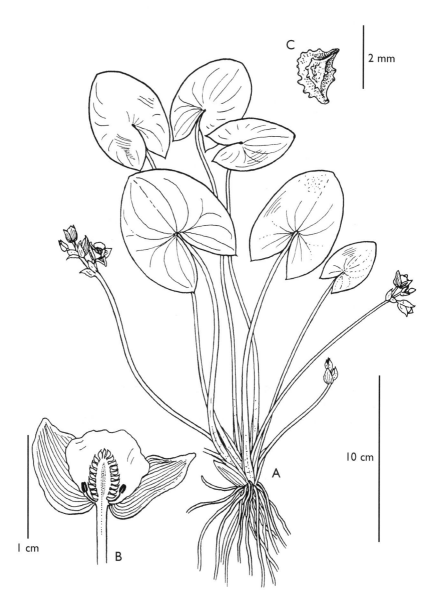

Fig. 21. *Sagittaria guayanensis* Kunth subsp. *guayanensis*: A, habit; B, flower with one side cut away; C, fruit with echinate margins. (Reproduced with permission from Haynes & Holm-Nielsen 1994).

Herb, perennial, to 50 cm tall, glabrous or sparsely pubescent; corms ca. 2.5 x ca. 3 cm. Leaves submersed or floating; submersed leaves sessile, pale green to greenish-brown, phyllodial, 0.4-0.7 x 0.4-0.8 cm, apex round acute to obtuse, basal lobes present, veins 3-5; floating leaves petiolate, petiole triangular, to 42 x 0.05-0.6 cm, basal sheath to 10 cm long, blade pale green, sagittate to hastate, 3.5-10.5 x 1.5-8.5 cm, apex round-acute, base cordate, sagittate, or hastate, basal lobes present, tips separated by 0.7-6.5 cm, veins 11-13. Inflorescence a simple scape of 1-7 whorls, floating, 0.9-9 x 0.6-4 cm, whorls with 3 flowers; peduncle terete, 10.9-22 x 0.09-0.5 cm; staminate bracts separate, delicate, elliptic to linear, 2.8-5.5 mm long, apex round, acute to obtuse; carpellate bracts separate, delicate, broadly elliptic, 9.4-15 mm long, apex obtuse; staminate pedicels spreading to erect, glabrous to densely pubescent, cylindric, 0.7-1.8 x 0.04-0.12 cm; carpellate pedicels erect to spreading in flower, spreading to reflexed in fruit, glabrous or rarely pubescent, cylindric, 0.6-2.1 x 0.12-0.21 cm. Staminate flowers: sepals erect, 5.5-10 x 3.8-5.2 mm, petals clawed, 6-8.4 x 4-6 mm, stamens 6, filaments glabrous, cylindric, ca. 1.5 x ca. 0.2 mm, anthers linear, 0.6-1.5 x ca. 0.3 mm, apex round acute, sterile carpels present. Carpellate flowers: sepals erect in flower and fruit, 4.5-10 x 2.5-10 mm, petals clawed, 8-10 x 5-7 mm, with ring of sterile stamens. Fruit aggregate 0.5-1.8 cm diam.; achenes obelliptic, 1.7-2.2 x 1.2-1.5 mm, tuberculate, without glands, with 1-3 lateral wings, margin echinate, beak lateral, erect, 0.2-0.5 mm long.

Distribution: In marshes, pools, and lakes, from southeastern U.S.A., Jamaica, Dominican Republic, and C Mexico S throughout C America and S America to Paraguay and northern Argentina; over 100 collections examined, including 22 from the Guianas (GU: 15; SU: 5; FG: 2).

Selected specimens: Guyana: Mazaruni-Potaro Region, Bartica, Linder 148 (GH, NY); Rupununi, basin of Rupununi R., A.C. Smith 2289 (A, G, NY, US, U). Suriname: Para Distr., Rijsdijkweg, Lindeman & Teunissen LBB 15219 (NY, U); Nickerie Distr., Nieuw-Nickerie, Hekking 939 (A, NY, U). French Guiana: Mana, Sagot 547 (BM).

Notes: The holotype of *S. guayanensis* in P is lost, the isotype is an insufficient fragment in MO; Rataj (1972: 10) selected Hostmann 870 as a neotype, which is, according the nomenclatural rules, an epitype.
The Paleotropic *S. guayanensis* subsp. *lappula* (D. Don) Bogin has the achenes compressed and more than 2.5 mm long.

2. **Sagittaria lancifolia** L., Syst. Nat. ed. 10. 1270. 1759. Type: Jamaica, Browne s.n., Herb. Linnaeus No. 1124/6 (lectotype LINN) (designated by Haynes & Holm-Nielsen 1994: 89). – Fig. 22

In the Guianas only: subsp. **lancifolia**

Sagittaria pugioniformis L., Pl. Surin. 15. 1775. Type: Suriname, Dahlberg s.n., Herb. Linnaeus No. 1124/7 (lectotype LINN) (designated by Haynes & Holm-Nielsen 1994: 90).
Sagittaria angustifolia Lindl., Bot. Reg. 14: t. 1141. 1828. – *Sagittaria lancifolia* Lindl. var. *angustifolia* (Lindl.) Griseb., Cat. Pl. Cub. 218. 1866. Type: [icon] Lindl., Bot. Reg. 14: t. 1141. 1828 (lectotype designated by Haynes & Holm-Nielsen 1994: 90).

Herb, perennial, to 2 m tall, glabrous to sparsely pubescent; rhizomes to 19 x to 4 cm. Leaves emersed; petioles round to triangular, 44-58 x 0.3-2.5 cm, basal sheath to 26 cm long, blade pale green, linear to ovate or elliptic, 20.3-35 x 0.7-16 cm, apex acute, basal lobes absent, veins 7-9. Inflorescence a simple or paniculate scape of 6-13 whorls, emersed, 16-72 x 19-60 cm, whorls with 3 flowers; peduncle triangular, 75-125.5 x 0.3-1.5 cm; bracts united, coarse, striate to ribbed, staminate bracts elliptic to ovate-elliptic, 3-6 mm long, apex round-acute to acute, carpellate bracts elliptic, 4-32 mm long, apex acute; staminate pedicels spreading, cylindric, 1.9-2.5 x 0.01-0.03 cm; carpellate pedicels spreading, in flower and fruit, to 38 x 0.3-0.4 mm. Staminate flowers: sepals reflexed to spreading, 5.5-6.5 x 3.5-5.2 mm, striate to ribbed, petals clawed, 0.8-1.5 x 0.6-1.2 mm, claws ca. 0.5 mm long, stamens ca. 24, filaments pubescent, cylindric, 2.2-4 x ca. 0.5 mm, anthers linear, 1.7-2.3 x ca. 0.8 mm, apex obtuse, without sterile carpels. Carpellate flowers: sepals reflexed to spreading in flower and fruit, 4-8.4 x 3.4-5.4 mm, striate to ribbed, petal clawed, 0.6-1.5 x 0.4-1.3 cm, claw to 4.5 mm long, without ring of sterile stamens. Fruit aggregate 0.5-1.2 cm diam.; achenes obelliptic, 1.6-2.5 x 0.8-1.1 mm, not tuberculate, with 1 gland, keeled, beak lateral, erect, 0.3-0.7 mm long.

Distribution: Coastal southeastern U.S.A., Caribbean Islands, Mexico, C America, and coastal S America from Colombia, Ecuador to Venezuela and C Brazil; more than 70 specimens examined, including 34 from the Guianas (GU: 22; SU: 8; FG: 4).

Fig. 22. *Sagittaria lancifolia* L.: A, bract of subsp. *media* (Micheli) Bogin, illustrating papillate veins; B, habit; C, bract of subsp. *lancifolia* illustrating non-papillate veins; D, stamen, illustrating pubescent filament; E, keeled fruit with one gland. (Reproduced with permission from Haynes & Holm-Nielsen 1994).

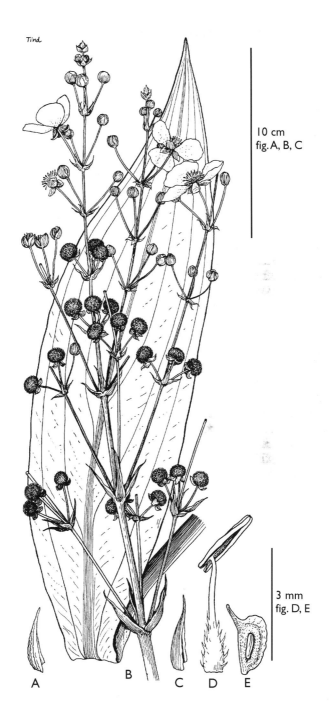

Tind.

10 cm
fig. A, B, C

3 mm
fig. D, E

A   B   C   D   E

142

Selected specimens: Guyana: Georgetown, Hitchcock 16887 (GH, NY, S); Mazaruni-Potaro Region, Kamakusa, de la Cruz 4097 (F, NY, US). Suriname: Perica, Wullschlaegel 493 (BR, GOET), 1566 (BR, GOET, U); without locality, Voltz s.n. (U). French Guiana: Route de St. Laurent à Mana, Feuillet 779 (U, UNA); Mana, Savane Cr. Jacques, W of Mana, Cowan 38878 (NY, US).

Economic use: Sold as aquarium plants under the German commercial name "Lanzettblättriges Pfeilkraut" ("Lance-leaved Arrowhead").

3. **Sagittaria rhombifolia** Cham., Linnaea 10: 219. 1835. Type: Brazil, Sellow s.n. (lectotype E, isolectotype K, LE) (designated by Rataj 1972: 14). – Fig. 23

Herb, perennial, to 100 cm tall, glabrous; rhizomes to 6 x to 4 cm wide; corms 1-4 x 0.6-2 cm. Leaves submersed or emersed; submersed leaves sessile, pale green, linear, 1-5 x 0.1-0.3 cm, apex acute, basal lobes absent, veins 1-3; emersed leaves petiolate, petiole more or less triangular, 13.5-68 x 0.6-1.5 cm, basal sheath to 23 cm long, blade pale green, linear-elliptic, elliptic-ovate, rhombiform, to subcordate, 8.5-20 x 0.2-15 cm, apex acute, basal lobes absent, base acute to rarely sub-cordate, veins 9-13. Inflorescence a simple scape of 2-10 whorls, emersed, 3.5-37 x 1-5 cm, whorls with 2-3 flowers; peduncle terete, 12-79 x 0.1-0.9 cm; staminate bracts united at base, delicate to coarse, elliptic, 6-22 mm long, apex round-acute to obtuse; carpellate bracts united at base, coarse, elliptic, 6-35 mm long, apex obtuse; staminate pedicels spreading, cylindric, 2.2-4.9 x 0.02-0.12 cm; carpellate pedicels erect in flower, spreading in fruit, cylindric in flower, becoming expanded in fruit, 0.5-2.5 x 0.2-0.55 cm. Staminate flowers: sepals erect, 9-11.5 x 4-6 mm, petals clawed, to 20.5 x to 13 mm, claws to 7 mm long, stamens 9-12, filaments glabrous, dilated, 2.3-3.8 x ca. 0.4 mm, anthers linear, 1-1.8 x 0.3-0.5 mm, apex round-acute, sterile carpels present. Carpellate flowers: sepals appressed in flower and fruit, nearly enclosing fruiting aggregate, 1.4-2.3 x 0.9-2.5 cm, petals white, yellowish, or pinkish clawed, 15-30 x 1.3-2.5 mm, claw ca. 5 mm long, without ring of sterile stamens. Fruit aggregate 1.2-3 cm diam.; achenes obelliptic, 3.2-7 x 2-3 mm, not tuberculate, without glands, keeled, beak lateral, horizontal, 0.7-1.2 mm long.

Fig. 23. *Sagittaria rhombifolia* Cham.: A, habit, illustrating leaves without expanded blades B, habit, illustrating leaves with expanded blades; C, stamen, illustrating glabrous filament; E, fruit without glands. (Reproduced with permission from Haynes & Holm-Nielsen 1994).

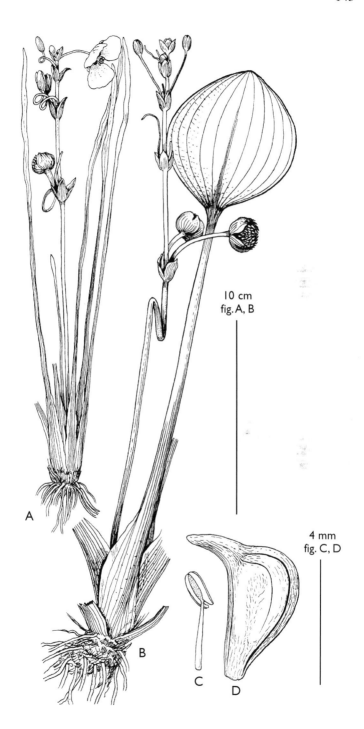

10 cm
fig. A, B

4 mm
fig. C, D

A

B

C

D

Distribution: In marshes and along margins of lakes and streams, from Costa Rica S to Argentina; more than 100 specimens examined, with 24 from the Guianas (GU: 22; SU: 2).

Selected specimens: Guyana: Georgetown, Hitchcock 16997 (GH, S, US); Rupununi Distr., Manari, Maas & Westra 3768 (NY, U). Suriname: 30 km S of Paramaribo, Teunissen LBB 14634 (U); Sipaliwini Savanna, Oldenburger et al. 1079 (U).

WOOD AND TIMBER

CYRILLACEAE

by

IMOGEN POOLE[11, 12, 13] & JIFKE KOEK-NOORMAN[13]

WOOD ANATOMY

The CYRILLACEAE forms a family of small shrubs and trees belonging to 3 genera of which only one, the monospecific *Cyrilla*, is found in the Guianas. *Cyrilla racemiflora* L. is a shrub, with many slender stems, or small tree. In tropical regions it can attain heights of 10 m or more (Record & Hess 1943; Thomas 1960). It grows close to water bodies thus accounting for the spongy, pliable bark at the base of the trunk. Although the timber has a fine uniform texture and is easy to work, its tendency to warp renders it with little commercial value however it has been exploited for charcoal production (Record & Hess 1943). Its bark is rich in phenolic compounds and has been used as a styptic or astringent (Thomas 1960 and references therein).

Terms are used in accordance with the defined descriptions according to the IAWA list of microscopic features for hardwood identification (IAWA Committee 1989).

GENERIC DESCRIPTION

CYRILLA Garden ex L. – Figs. 24, 25

Growth rings indistinct or absent.
Vessels diffuse, solitary, paired, rarely groups in of 3, 68-114 (92) per mm^2, circular but predominantly angular, 29-73 (52) µm wide. Perforation plates scalariform with many (20-40, rarely more) fine closely spaced

[11] Department of Molecular Palaeontology, Faculty of Earth Sciences, Utrecht University, P.O. Box 80021, 3508 TA Utrecht, The Netherlands.
[12] Palaeoecology, Institute of Environmental Biology, Faculty of Science, Utrecht University, Laboratory of Palaeobotany and Palynology, Budapestlaan 4, 3584 CD Utrecht, The Netherlands.
[13] National Herbarium of the Netherlands, Utrecht University branch, Heidelberglaan 2, 3584 CS Utrecht, The Netherlands.

Acknowledgements: We are grateful to Lubbert Westra for providing the hand lens photograph of *Cyrilla racemiflora*. This work was funded by NWO/ALW Grant number 815.01.005.

146

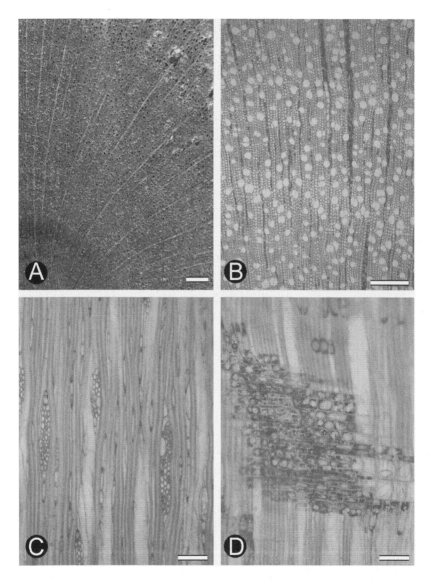

Fig. 24. *Cyrilla racemiflora* L.: A, transverse section hand lens view, scale bar = 500 μm, (Uw 27344); B, transverse section, scale bar = 250 μm (Uw 27330); C, tangential section, scale bar = 100 μm (Uw 27330); D, radial section, scale bar = 100 μm (Uw 27330).

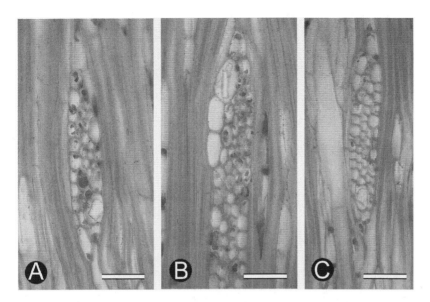

Fig. 25. *Cyrilla racemiflora* L.: A-C, tangential sections, scale bar = 100 μm (Uw 27330).

bars. Intervessel pits and vessel-ray pits opposite, elongated and scalariform with distinct borders.
Rays uniseriate and multiseriate (3-6 cells), 10-14 per mm. Uniseriate rays composed of upright cells. Multiseriate rays heterogeneous composed predominantly of procumbent and square body ray cells and 1-3 rows of (square to) upright marginal cells. Multiseriate rays 23-83 (43) μm wide and 225-1475 (695) μm in height. Prismatic crystals abundant in the ray cells.
Axial parenchyma scanty and diffuse; 3-8 cells per strand.
Ground tissue of non-septate fibres with thin-to-thick walls, and bordered pits on both the radial and tangential walls.

N o t e : Growth ring boundaries can be distinct and the wood can also be semi-ring-porous according to the InsideWood database (2004-onwards) and Araujo & Mattos Filho (1974). Gummy deposits common in vessels according to Araujo & Mattos Filho (1974). Various types of pit membrane remnants associated with the perforation plates have been observed in *C. racemiflora* by Schneider & Carlquist (2003).

M a t e r i a l   s t u d i e d (Uw-numbers refer to the Utrecht Wood collection): **Cyrilla racemiflora** L.: Guyana: Maas *et al.* 5752 (Uw 27344); Maas *et al.* 5701 (Uw 27330).

THEOPHRASTACEAE

by

FREDERIC LENS[14] & STEVEN JANSEN[15]

WOOD ANATOMY

FAMILY CHARACTERISTICS

The wood of THEOPHRASTACEAE is characterised by narrow (often less than 70 μm) and short vessel elements (commonly less than 500 μm in length); vessels are arranged solitary or in radial multiples of 2-7 cells in *Clavija* and *Theophrasta* L., while vessel clustering is observed in *Deherainia* Decne. and especially in *Bonellia* Bertero ex Colla and *Jacquinia* L. Vessel perforations are exclusively simple, and intervascular pitting is small (3-5 μm in horizontal width) and alternate. Vessel density ranges from 20/mm^2 in *Bonellia* and *Jacquinia* to 160/mm^2 in *Clavija*. Axial parenchyma is paratracheal and sparse. Rays are exclusively multiseriate consisting of procumbent and square cells in *Clavija* and *Deherainia*, but mainly procumbent ray cells in *Bonellia*, *Jacquinia* and *Theophrasta*. The rays are often 6-10-seriate (up to 36-seriate) and 200-900 μm wide, and usually between 2000-6000 μm high. Fibres are short (often less than 700 μm in length), septate only in *Clavija* and *Theophrasta*, thin- to thick-walled, and they have simple to minutely bordered pits concentrated in the radial walls. Small calcium oxalate crystals of various shapes found in rays of *Jacquinia*; large spherical clusters of needle-like crystals in rays of *Theophrasta*. Silica grains in ray cells are present in *Clavija* and *Bonellia*.

The generic description is based on 8 wood samples of *Clavija*. As far as we know, the wood is of no commercial value (Record & Hess 1936, p. 28).

GENERIC DESCRIPTION

CLAVIJA Ruiz & Pav.   – Figs. 26, 27

Vessels round to oval, solitary and in short radial multiples of 2-3, diffuse-porous (Figs. 26 A, 27 A); diameter 12-40 μm; 40-100 vessels per mm^2. Perforations simple (Fig. 27 B). Intervascular pits alternate, 4-5 μm. Vessel-ray pits similar to intervessel pitting. Vessel member length 260-570 μm.

---

[14] Laboratory of Plant Systematics, Institute of Botany and Microbiology, K.U. Leuven, Kasteelpark Arenberg 31, BE-3001 Leuven, Belgium.

[15] Jodrell Laboratory, Royal Botanic Gardens, Kew, Richmond, Surrey TW9 3DS, U.K.

149

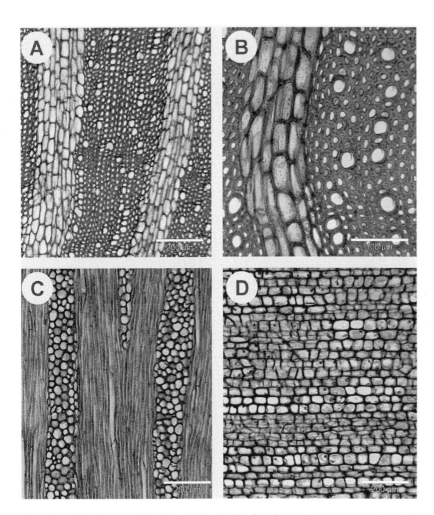

Fig. 26. LM pictures: *Clavija lancifolia* Desf. subsp. *chermontiana* (Standl.)
B. Ståhl: A, transverse section (Uw 36488); B, transverse section (Uw 32076);
C, tangential section. (Uw 32076); D, radial section (Uw 10934).

Rays exclusively multiseriate, 3-14 cells wide (Fig. 26 C), composed of
procumbent and square body ray cells and 1 to 2 square to upright
marginal ray cells (Fig. 26 D); height 750-8700 µm, density 1-3 per mm.
Axial parenchyma scarce and scanty paratracheal as 2-4 celled strands;
strands 350-550 µm high.

150

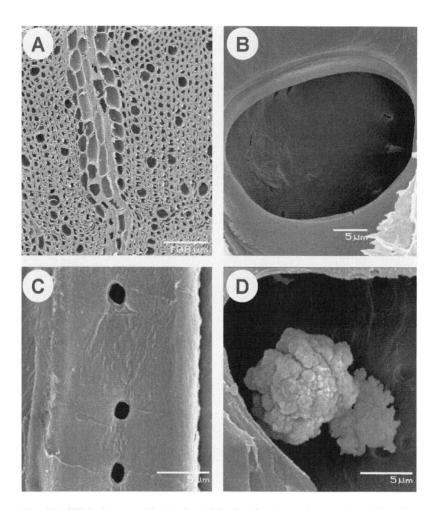

Fig. 27. SEM pictures: *Clavija lancifolia* Desf. subsp. *chermontiana* (Standl.) B. Ståhl: A, transverse section (Uw 10934); B, radial section showing simple vessel perforation plate. (Uw 10934); C, radial section showing simple to minutely bordered fibre pits (Uw 32076); D, radial section showing silica body in ray cell (Uw 30599).

Fibres septate and thin- to thick-walled (Fig. 26 B), walls 2-4 μm, lumina 7-15 μm wide. Pits simple or sometimes minutely bordered (Fig. 27 C), 2-3 μm in size, oval to slit-like, concentrated in radial walls.
Silica bodies present in ray cells (Fig. 27 D).
Groups of sclereids in pith.

The wood of *Clavija lancifolia* corresponds with other *Clavija* species studied, and shows many similarities with *Theophrasta*. Both genera differ from *Bonellia*, *Deherainia* and *Jacquinia* based on their narrower vessels (generally less than 30 µm in tangential width), higher vessel densities (often between 60-80 per mm^2), longer vessel elements (usually 300-450 µm) and septate fibres, while the latter three genera can be identified by their pronounced vessel clustering (Lens *et al.* 2005).

Material studied (Uw-numbers refer to the Utrecht Wood collection):

Clavija lancifolia Desf. subsp. **lancifolia**: Suriname: Lindeman & Görts *et al.* 518 (Uw 26481; diam. 0.75 cm); Lindeman 4463 (Uw 3122; diam. 3.2 cm). French Guiana: de Granville *et al.* 6509 (Uw 30000; diam. 2 cm).

C. lancifolia Desf. subsp. **chermontiana** (Standl.) B. Ståhl: Guyana: Jansen-Jacobs *et al.* 583 (Uw 32076; diam. 1 cm); Jansen-Jacobs *et al.* 1518 (Uw 33202; diam. 1.6 cm); Jansen-Jacobs *et al.* 5634 (Uw 36488; diam. 1.5 cm). Suriname: Florschütz & Maas 2475 (Uw 10934; diam. 2.5 cm); Lindeman 6693 (Uw 4524; diam. 3 cm).

C. macrophylla (Link ex Roem. & Schult.) Miq.: Guyana: Jansen *et al.* 356 (Uw 30599; diam. 2.5 cm).

RHABDODENDRACEAE

by

IMOGEN POOLE[16, 17, 18] & JIFKE KOEK-NOORMAN[18]

WOOD ANATOMY

The monogeneric family RHABDODENDRACEAE comprises 3 species of shrubs or small trees confined to tropical South America (Prance 1972). In the Guianas, RHABDODENDRACEAE is represented by one arboreal species, *Rhabdodendron amazonicum* (Spruce ex Benth.) Huber, with no commercial value.

Terms are used in accordance with the defined descriptions according to the IAWA list of microscopic features for hardwood identification (IAWA Committee 1989).

GENERIC DESCRIPTION

RHABDODENDRON Gilg & Pilg. – Fig. 28

Successive bundles of xylem and concentric included phloem present alternating with bands of conjunctive tissue (Fig. 28 A). Growth rings absent.

Vessels diffuse, mostly solitary, 3-11 (7) per mm^2, generally more abundant in the vicinity of the conjunctive tissue, circular to oval (Fig. 28 B) and medium in size with tangential diameters ranging from 44-171 (107) μm wide. Perforation plates simple. Intervessel pits minute, with slit-like pit apertures, vestured, alternately arranged. Vessel-ray pits bordered, similar in size and arrangement to intervessel pits.

Rays of 2 sizes, 1-3-seriate and (less frequently) 4-7-seriate (Fig. 28 C), 6-13 per mm. Uniseriate rays homocellular composed of square to

[16] Department of Molecular Palaeontology, Faculty of Earth Sciences, Utrecht University, P.O. Box 80021, 3508 TA Utrecht, The Netherlands.

[17] Palaeoecology, Institute of Environmental Biology, Faculty of Science, Utrecht University, Laboratory of Palaeobotany and Palynology, Budapestlaan 4, 3584 CD Utrecht, The Netherlands.

[18] National Herbarium of the Netherlands, Utrecht University branch, Heidelberglaan 2, 3584 CS Utrecht, The Netherlands.

Acknowledgements: We are grateful to Lubbert Westra for providing the hand lens photograph of *Rhabdodendron amazonicum*. This work was funded by NWO/ALW Grant number 815.01.005.

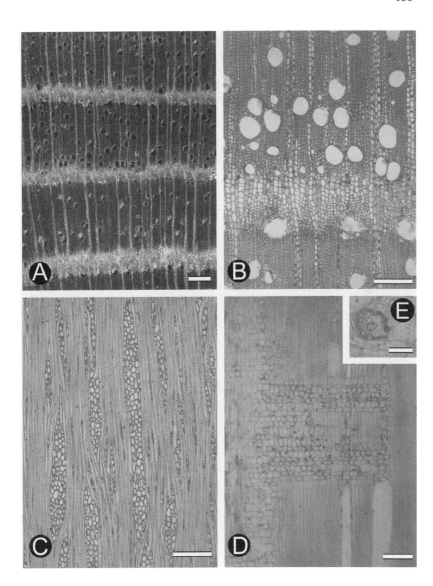

Fig. 28. *Rhabdodendron amazonicum* (Spruce ex Benth.) Huber (Uw 11162): A, transverse section hand lens view, scale bar = 500 μm; B, transverse section, scale bar = 250 μm; C, tangential section, scale bar = 250 μm; D, radial section, scale bar = 250 μm; E, radial section, scale bar = 25 μm.

upright cells. Multiseriate rays heterogeneous composed predominantly of square and upright body cells and 1-3(-6) rows of upright marginal cells (Fig. 28 D). Multiseriate rays 28-184 (71) μm wide and 263- >3000 (1200) μm high, rarely composed. Amorphous contents present in most ray cells. Sheath cells occasionally present. Perforated ray cells rarely present. Rare prismatic crystals or druses present in the ray body cells (Fig. 28 E).

Parenchyma as conjunctive tissue forming strands of more than 10 cells; otherwise axial parenchyma sparsely diffuse and scanty paratracheal with axial strands of 2-3 cells.

Ground tissue of non-septate fibre-tracheids with thick walls and bordered pits on both radial and tangential walls.

Note: Bordered pits on the fibre tracheids are vestured according to Carlquist (2001). Successive cambium not present in twig wood (Heimsch 1942; Prance 1968, 1972).

Material studied (Uw-numbers refer to the Utrecht Wood collection): **Rhabdodendron amazonicum** (Spruce ex Benth.) Huber: Suriname: LBB 10694 (Uw 11162); Schulz 8322a (Uw 6800).

PROTEACEAE

by

IMOGEN POOLE[19, 20, 21] & JIFKE KOEK-NOORMAN[21]

WOOD ANATOMY

The S American PROTEACEAE comprise small shrubs and trees belonging to about 60 genera within the tribe Macadamieae, of which only 3 genera, *Euplassa* Salisb. ex Knight, *Panopsis* Salisb. ex Knight and *Roupala* Aubl., are found in the Guianas. These 3 genera, each with over 20 species, occur predominantly in the highlands with their centres of species diversity occurring in the Andes and Guayana Highlands. Therefore it has been suggested that *Roupala* and *Panopsis* are today not true rainforest species and only occupy refugia in the lowlands as their preference lies in the more seasonal environments offered by higher altitudes (Prance & Plana 1998). *Euplassa* is a S American genus of trees, rarely shrubs, with 2 species occurring in the Guianas. Two species of *Panopsis* are native to the Guianas whereas *Roupala* comprises the largest genus of the tropical American PROTEACEAE with about 32 species (Edwards & Prance 2003) of which only 5 species occur in the Guianas. According to Mennega (1966) the wood of *Euplassa* and *Panopsis* are very similar whereas *Roupala* is more distinct. All 3 genera share the characteristic feature of wide rays that can extend several millimetres and are conspicuous in all three planes of section (Mennega 1966).

GENERIC DESCRIPTIONS

EUPLASSA Salisb. ex Knight   – Fig. 29 A-F

Growth rings indistinct or absent.

[19] Department of Molecular Palaeontology, Faculty of Earth Sciences, Utrecht University, P.O. Box 80021, 3508 TA Utrecht, The Netherlands.
[20] Palaeoecology, Institute of Environmental Biology, Faculty of Science, Utrecht University, Laboratory of Palaeobotany and Palynology, Budapestlaan 4, 3584 CD Utrecht, The Netherlands.
[21] National Herbarium of the Netherlands, Utrecht University branch, Heidelberglaan 2, 3584 CS Utrecht, The Netherlands.

Acknowledgements: We are grateful to Lubbert Westra for providing the hand lens photographs of the transverse sections of *Euplassa, Panopsis* and *Roupala*. This work was funded by NWO/ALW Grant number 815.01.005.

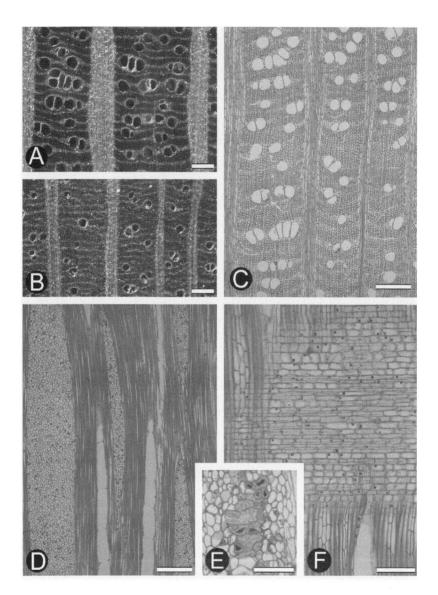

Fig. 29. *Euplassa pinnata* (Lam.) I.M. Johnst.: A, transverse section hand lens view, scale bar = 500 μm (Uw 5697); B, transverse section hand lens view, scale bar = 500 μm (Uw 1905); C, transverse section, scale bar = 1 mm (Uw 5697); D, tangential section, scale bar = 1 mm (Uw 5754); E, tangential section, scale bar = 200 μm (Uw 1905); F, radial section, scale bar = 250 μm (Uw 5697).

Vessels diffuse, solitary, paired or in groups of 3-4, rarely more, which tend to be tangentially arranged and occasionally appear as tangential bands; 3-14 (9) per mm², circular 45-258 (148) µm wide. Perforation plates simple. Intervessel pits alternate and usually bordered. Vessel-ray pits alternate.

Rays uniseriate and multseriate (5-22 cells wide), 1-2 per mm. Uniseriate rays composed of square to upright cells and 1-2 cells in height. Multiseriate rays composed of predominantly procumbent body cells and 1-3 (extremely rarely up to 5) rows of upright and/or square marginal cells. Multiseriate rays 78-1100 (540) µm wide and over 1 mm in height, sometimes appearing as an aggregate, dissected by axial elements. Nested sclerotic ray parenchyma cells (Fig. 29 E) can be present in the multiseriate rays along with amorphous (non crystalline) deposits.

Axial parenchyma present in narrow tangential bands 1-2 cells wide and 4-8 per mm, with 2 to many cells per strand. Ground tissue of non septate fibres with thin to thick walls and rare simple to minutely bordered pits on the radial walls.

N o t e : *Euplassa* is also described as having bordered pits on the fibres (Détienne & Jacquet 1983). Fibre pits are also noted on tangential walls equal in abundance to those on the radial walls (Mennega 1966).

M a t e r i a l   s t u d i e d (Uw-numbers refer to the Utrecht Wood collection): **Euplassa pinnata** (Lam.) I.M. Johnst.: Suriname: Lindeman 7607 (Uw 4722); Lanjouw & Lindeman 2759 (Uw 1905). French Guiana: BAFOG 1207 (Uw 5697); BAFOG 1266 (Uw 5754).

PANOPSIS Salisb. ex Knight — Fig. 30 A-E

Growth rings indistinct or absent.
Vessels diffuse, solitary, paired or in groups of 3-4, rarely more, which can be either radially or tangentially arranged and occasionally appear as tangential bands in *P. rubescens*; 2-13 (6) per mm², circular 35-325 (143) µm wide. Perforation plates simple. Intervessel pits alternate and bordered. Vessel-ray pits alternate and bordered.

Rays uniseriate and multseriate (4-20 cells wide), 1-3 per mm. Uniseriate rays composed of upright or square to upright cells and 2-20 cells in height. Multiseriate rays composed of predominantly procumbent body cells and 1-4(-6) rows of square to upright marginal cells. Multiseriate rays 62-788 (427) µm, sometimes >1 mm, wide and 29 µm to >1 mm in height, sometimes appearing as an aggregate, dissected by axial elements. Amorphous (non crystalline) deposits can be present in the multiseriate rays along with occasional silica bodies.

158

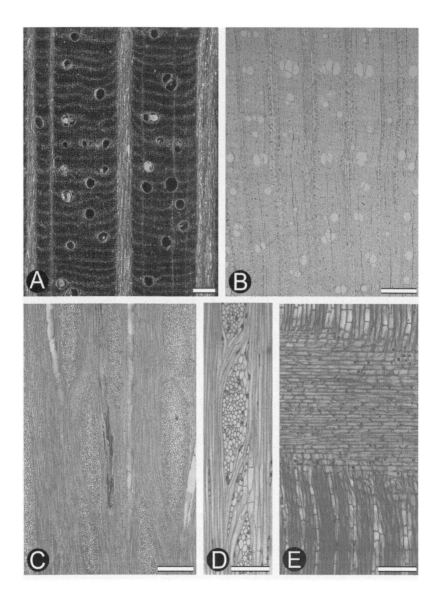

Fig. 30. A-B, D-E: *Panopsis sessilifolia* (Rich.) Sandwith: A, transverse section hand lens view, scale bar = 500 μm *(*Uw 5810); B, transverse section, scale bar = 1 mm (Uw 4399); D, tangential section, scale bar = 250 μm (Uw 4399); E, radial section, scale bar = 250 μm (Uw 4399). C: *Panopsis rubrescens* (Pohl) Pittier: C, transverse section, scale bar = 1 mm (Uw 32287).

Axial parenchyma present in narrow tangential bands 1-2(3) cells wide and 3-5 per mm; with 2 to many cells per strand. Ground tissue of non septate fibres with thin to thick walls and rare simple to minutely bordered pits on the radial walls.

Note: Vessel diameter is recorded as being up to 800 μm in the InsideWood database and narrower by Détienne & Jacquet (1983). *Panopsis* is described as having bordered pits on the fibres (Détienne & Jacquet 1983). Fibre pits are also noted on tangential walls equal in abundance to those on the radial walls (Mennega 1966).

Material studied (Uw-numbers refer to the Utrecht Wood collection): **Panopsis rubrescens** (Pohl) Pittier: Brazil: Krukoff 7236 (Uw 8264). Guyana: Jansen-Jacobs *et al.* 1357 (Uw 32287). Suriname: Lanjouw & Lindeman 2872 (Uw 1960).
P. sessilifolia (Rich.) Sandwith: Guyana: Forest Dept. British Guiana 3040 (Uw 1039). Suriname: Lindeman 6413 (Uw 4399); Stahel 291 (Uw 291). French Guiana: BAFOG 1322 (Uw 5810).

ROUPALA Aubl.   – Fig. 31 A-F

Growth rings indistinct or absent.
Vessels diffuse, solitary and clustered into groups of up to 3, rarely more forming tangential bands; 13-19 (16) per mm^2, circular, 38-128 (84) μm wide. Perforation plates simple. Intervessel pits alternate and usually bordered. Vessel-ray pits alternate.
Rays uniseriate and multiseriate (6-26 cells wide), 1-4 per mm. Uniseriate rays composed of upright cells and 1-5 cells in height. Multiseriate rays composed of procumbent body cells and 1-3 rows of upright and/or square marginal cells. Multiseriate rays 175-1075 (493) μm wide and over 1mm in height, often appearing as an aggregate, dissected by axial elements. Sclerotic ray parenchyma cells present in the multiseriate rays of *R. suaveolens* whereas silica bodies present in the ray and axial parenchyma cells of *R. montana* (Fig. 31 E).
Axial parenchyma present in narrow tangential bands 1-3 cells wide in *R. suaveolens* and 2-5 cells wide in *R. montana* and 3-5 per mm; up to 10 cells per strand. Ground tissue of non septate fibres with thick walls and rare simple to minutely bordered pits on the radial walls.

160

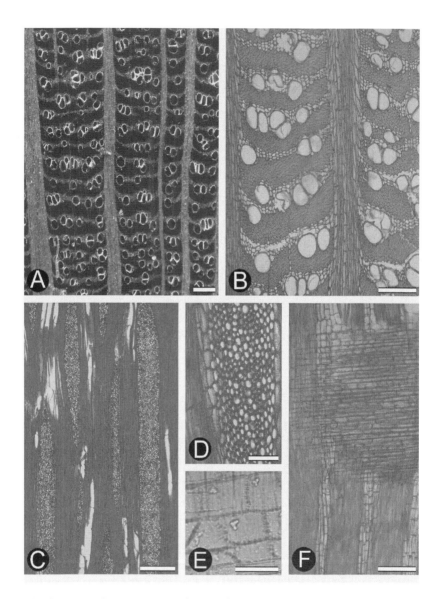

Fig. 31. *Roupala montana* Aubl. (Uw 249): A, transverse section hand lens view, scale bar = 500 µm; B, transverse section, scale bar = 1 mm; C, tangential section, scale bar = 1 mm; D, tangential section, scale bar = 100 µm; E, radial section, scale bar = 100 µm; F, radial section, scale bar = 250 µm.

N o t e : Multiseriate rays are wider in *R. montana* (up to 26 cells or 1075 µm) relative to *R. suaveolens* (up to 13 cells or 625 µm). Silica bodies have also been recorded in the axial parenchyma cells of *R. montana* according to the InsideWood database. *Roupala* is also described as having bordered pits on the fibres (Détienne & Jacquet 1983). Fibre pits are also noted on tangential walls but less numerous than those on the radial walls (Mennega 1966).

M a t e r i a l s t u d i e d (Uw-numbers refer to the Utrecht Wood collection): **Roupala montana** Aubl.: Suriname: Stahel 249 (Uw 249).

R. suaveolens Klotzsch: Guyana: Jansen-Jacobs *et al.* 1628 (Uw 33243).

COMBRETACEAE

by

Jifke Koek-Noorman[22], Imogen Poole[22, 23, 24],
Lubbert Y.T. Westra[22] & Jan W. Maas[22]

WOOD ANATOMY

INTRODUCTION

The most recent global treatment of the wood anatomy of the COMBRETACEAE is by van Vliet (1979). Therefore, his descriptions were used as a basis for this chapter, supplemented by our own observations. In van Vliet's thorough study, each generic description was followed by an extensive discussion of the individual characters, their taxonomical value and their correlation with habit and habitat, and the classification of the family. Most of those data and conclusions go beyond the scope of this Flora. For example: when studying the bordered intervessel pits in COMBRETACEAE and allied families, van Vliet (1978, 1979) was able to distinguish between 2 main types, based on size of the vestures and their position in the pit chamber. One intervessel pit type (type A) was characteristic for the genus *Strephonema* Hook.f. (not studied here). The other type (type B), found in all other genera of COMBRETACEAE, was subdivided into 3 forms, but these subtypes were not correlated with other woodanatomical characters, nor did van Vliet find any correlation with the taxonomy of the family. Only in *Terminalia* the full variation of type B was found. As the use of a scanning electron microscope is essential to decide on the type of vesturing in a given sample, we refrained from describing the vesture types here. Moreover, a recapitulation here of van Vliet's discussion of anatomical variation in relation to habit and habitat, but restricted to Guianan material only, would not contribute to, let alone improve, his conclusions and thus the reader is referred to the original work (van Vliet 1978).

Other quantitative characters that may be difficult to measure accurately from sections, such as vessel member length, ray height, are not given here, notwithstanding the fact they may have diagnostic value.

[22] National Herbarium of the Netherlands, Utrecht University branch, Heidelberglaan 2, 3584 CS Utrecht, The Netherlands.
[23] Department of Molecular Palaeontology, Faculty of Earth Sciences, Utrecht University, P.O. Box 80021, 3508 TA Utrecht, The Netherlands.
[24] Palaeoecology, Institute of Environmental Biology, Faculty of Science, Utrecht University, Laboratory of Palaeobotany and Palynology, Budapestlaan 4, 3584 CD Utrecht, The Netherlands.

Van Vliet reports 2 distinct types of vessel elements for *Combretum*. Besides the elements typical for COMBRETACEAE, very narrow elements, often associated with vascular tracheids are present. In the Guianan species of *Combretum*, however, they are rarely found.

Descriptive terms and measurements are in accordance to the IAWA-list of microscopic features (1989).

FAMILY DESCRIPTION

Growth rings absent to distinct.

Vessels diffuse, round to oval, solitary and in short radial multiples of 2-3(-10); typical vessels in *Combretum* rarely intermingled with very narrow vessels and vascular tracheids in small clusters; mean tangential diameter 87-170 μm; 3-25 per mm². Perforations simple with horizontal to oblique end walls. Intervessel pits vestured, alternate, round to polygonal, (5-)6-9(-11) μm, rarely coalescent; vessel-ray pits similar to intervessel pits but half-bordered, sometimes in horizontal arrangement, elongate, rarely tending to scalariform or coalescent.

Rays exclusively uniseriate in *Laguncularia*, nearly so in *Buchenavia*, *Combretum, Conocarpus* and *Terminalia*. Rays composed of square and upright, and weakly procumbent cells in *Buchenavia* and *Laguncularia*. In *Conocarpus* upright to procumbent cells are intermingled. In *Combretum* and *Terminalia*, cells are procumbent (with infrequent square and upright cells and upright marginal cells); 7-11 per mm, but up to 20 in *Combretum* and *Terminalia*. Perforated ray cells present in *Combretum*.

Parenchyma apotracheal diffuse, sometimes in marginal bands or absent; paratracheal vasicentric to aliform-confluent, sometimes in bands of 2-5 cells wide; strands of (3-)4-7(-8) cells.

Ground tissue fibres thin-walled to very thick-walled; non-septate in *Laguncularia*, but septate in *Buchenavia* and *Conocarpus*. In *Combretum* and *Terminalia*, both septate and non-septate fibres can be found; pits simple to minutely bordered, mainly on radial cell walls.

Crystals present in rays and/or axial parenchyma, as small or large rhombic crystals (Fig. 38), or as elongated rod- to styloid-like crystals, sometimes scanty (Table 1).

N o t e s : Five Guianan genera are represented, 2 of which by 1 species only. Due to the variation within the large genera *Combretum* and *Terminalia*, it is impossible to separate these genera. However, some characters can be used for a preliminary identification purposes. Noteworthy here is that the radial vessels reported by van Vliet (1979) could not be found in our material.

The parenchyma pattern is variable, even within species. According to van Vliet, the diagnostic value is restricted to those species "where distinctly different types (*e.g.* aliform and banded) are compared".

Table 1. Occurrence of crystals in Guianan COMBRETACEAE.
Data between brackets: crystals are present in some species.
?: crystals have been reported by van Vliet (1979), but not confirmed in this study.

	rays	axial parenchyma
Buchenavia	(rhombic, elongate)	
Combretum	small, rhombic, in large ideoblasts	large, rhombic, in large ideoblasts
Conocarpus	large, blunt prismatic, in radial rows	
Laguncularia	large, prismatic to elongate	
Terminalia	(large, blunt prismatic)	scanty rhombic; druses in *T.catappa*?

GENERIC DESCRIPTIONS

BUCHENAVIA Eichler   – Figs. 32 A-D; 33 A-B

Growth rings indistinct or absent.
Vessels diffuse, round to oval, predominantly solitary and in infrequent small radial multiples of 2(-3) cells, but up to 8 in *B. tetraphylla*; tangential diameter 90-170(-200) Ìm, often a few small vessels present, 3-6(-8) per mm², but up to 12 in *B. tetraphylla*. Perforations simple. Intervessel pits alternate, polygonal, 6-8 μm, vestured; vessel-ray pits similar to intervessel pits but half-bordered.
Rays uniseriate, consisting of procumbent body ray cells and 1 or 2 rows of square to upright marginal cells; cells square or nearly so in *B. macrophylla*, square and upright in *B. tetraphylla*; 5-11(-13) /mm.
Parenchyma apotracheal scantily diffuse; paratracheal vasicentric to aliform-confluent, often connected to narrow marginal bands; strands of 4-7 cells.

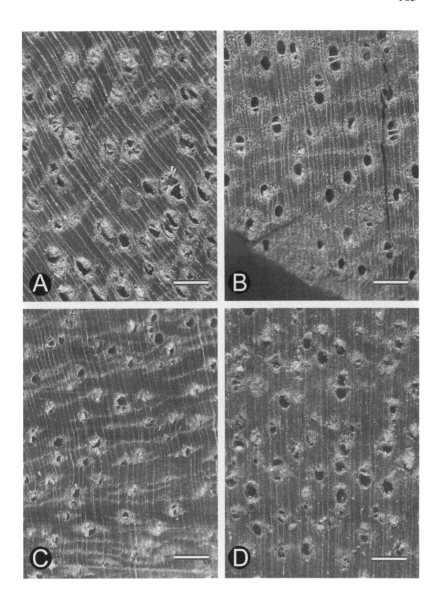

Fig. 32. A-D: *Buchenavia,* transverse sections, scale bar = 500 µm: A, *B. fanshawei* Exell & Maguire (Uw 34322); B, *B. grandis* Ducke (Uw 7931); C, *B. macrophylla* Eichler (Uw 7494); D, *B. tetraphylla* (Aubl.) R.A. Howard (Uw 686).

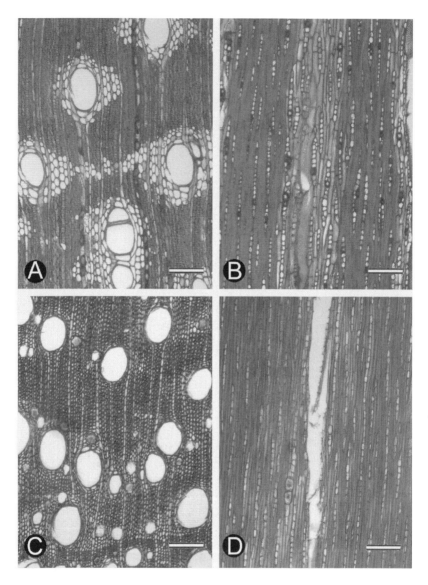

Fig. 33. A-B: *Buchenavia grandis* Ducke (Uw 7754), scale bar = 110 μm: A, transverse section; B, tangential section; C-D: *Combretum*, scale bar = 110 μm: C, *C. fruticosum* (Loefl.) Stuntz: transverse section (Uw 30503); D, *C. rotundifolium* Rich.: tangential section (Uw 1574).

Prismatic to elongate crystals sometimes present, filling the ray or parenchyma cells. Solid contents present in axial and ray parenchyma of *B. macrophylla*. Small granular silica grains reported by van Vliet (1979) for *B. acuminata* and *B. fanshawei*.
Ground tissue fibres thin- to thick-walled, or very thick-walled, partly septate but frequently so in *B. fanshawei*, pits simple to minutely bordered, mainly on radial walls. In *B. tetraphylla*, some bands of thin-walled septate fibres alternate with the thick-walled, partly septate fibres.

N o t e : The sample of *B. tetraphylla* deviates from the other species studied in a number of characters: the presence of infrequent radial vessel multiples, the square/upright ray cells, the dark solid contents in the rays and parenchyma cells, and the bands of thin-walled fibres.

M a t e r i a l   s t u d i e d (Uw-numbers refer to the Utrecht Wood collection): **Buchenavia fanshawei** Exell & Maguire: Guyana: Polak 576 (Uw 34322).
B. grandis Ducke: Brazil: Krukoff 6472 (Uw 7754); Krukoff 6794 (Uw 7931).
B. macrophylla Eichler: Brazil: Krukoff 6117 (Uw 7494).
B. tetraphylla (Aubl.) R.A. Howard: Suriname: BBS 91 (Uw 686).
B. ochroprumna Eichler: French Guiana: BAFOG = Bena 1313 (Uw 5801).

COMBRETUM Loefl.   – Figs. 33 C-D; 34 A-D

Growth rings faint to distinct.
Vessels diffuse, of 2 distinct sizes: normal vessels solitary, 11-15 per mm^2 on av., mean tangential diameter 115-126 μm, mixed with narrow vessels, mean tangential diameter 20-30 μm. Perforations simple. Intervessel pits alternate, round to polygonal, 5-7 μm but those of the very narrow vessels infrequently elongate, vestured; vessel-ray pits similar to intervessel pits but half-bordered, sometimes in distinct horizontal rows and coalescent.
Rays uniseriate, ca 10-20 (25) per mm, composed of square and (weakly) procumbent cells, with some rows of upright cells; upright cells more frequently present in *C. fruticosum* and *C. pyramidatum*. Dark contents in the rays of *C. cacoucia*.
Parenchyma apotracheal diffuse, in marginal bands and paratracheal, vasicentric-aliform; mainly aliform-confluent in *C. pyramidatum*; strands of 4-6(-8) cells.

168

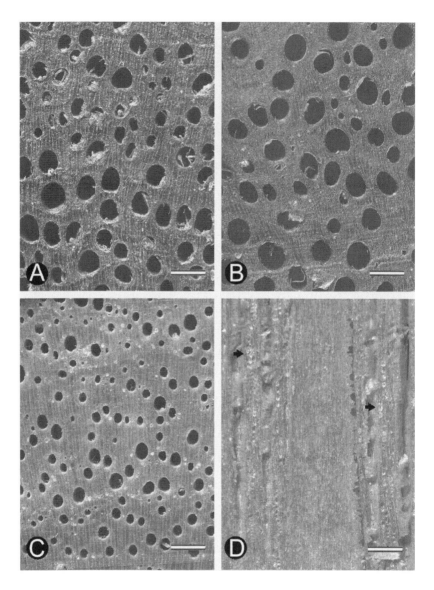

Fig. 34. A-D: *Combretum*, scale bar = 500 μm: A, *C. cacoucia* Exell: transverse section (Uw 33388); B, *C. laxum* Jacq.: transverse section (Uw 22691); C-D: *C. fruticosum* (Loefl.) Stuntz (Uw 30503): C, transverse section; D, tangential section, arrows pointing towards crystals in axial parenchyma strands.

Ground tissue fibres thin-walled and septate in *C. fruticosum*, thin- to thick-walled and partly septate in *C. pyramidatum*, thick to very thick-walled and non-septate in *C. rotundifolium*; frequently gelatinous. Pits simple to minutely bordered, mainly on radial cell walls.
Crystals in all species studied. Enlarged cells containing one large solitary crystal in the rays and in few axial parenchyma strands of *C. fruticosum*. Small to large rhombic crystals in ray cells of *C. cacoucia* and *C. laxum*, rhombic crystals filling the cells in axial parenchyma strands of *C. rotundifolium*.

N o t e : In the Guianas 9 species occur, 5 of which are represented in this study. Van Vliet (1979) published a very detailed wood anatomical description of *Combretum*, based on material of 14 species, 3 of which are from the Neotropics. The variation he reports is considerable, particularly in mean quantitative characters such as percentage of solitary vessels, vessel frequency and diameter, and the presence of very narrow vessels. However, the variation was mainly due to African representatives. Although included phloem was found in part of the African species, it was absent from the material from the Neotropics. Likewise, radial vessels, were seemingly absent.

M a t e r i a l   s t u d i e d (Uw-numbers refer to the Utrecht Wood collection):
Combretum cacoucia Exell: Guyana: Ursem & Potters 34 (Uw 33388).
C. fruticosum (Loefl.) Stuntz: Guyana: Jansen-Jacobs *et al.* 149 (Uw 30503).
C. laxum Jacq.: Suriname: Heyde & Lindeman 24 (Uw 22691).
C. pyramidatum Ham.: Suriname: Lanjouw & Lindeman 1861 (Uw 1565).
C. rotundifolium Rich.: Suriname: Lanjouw & Lindeman 1873 (Uw 1574).

CONOCARPUS L.   – Fig. 35 A-B

Growth rings faint to distinct.
Vessels diffuse, round to oval, solitary and in small radial multiples of 2-3(-5), tangential diameter (38-)87-(120) μm, (17-)25(-33) per mm². Perforations simple. Intervessel pits alternate, round to polygonal, 4-5 μm but infrequently elongate up to 10 μm, vestured; vessel-ray pits similar to intervessel pits but half-bordered, sometimes elongate to 15 μm. Sometimes deposits in heartwood.
Rays uniseriate with scanty biseriate parts, (5-)7(-9) per mm, composed of upright to procumbent cells. Large solitary crystals frequent in idioblastic ray cells, these cells in radial arrangement.

170

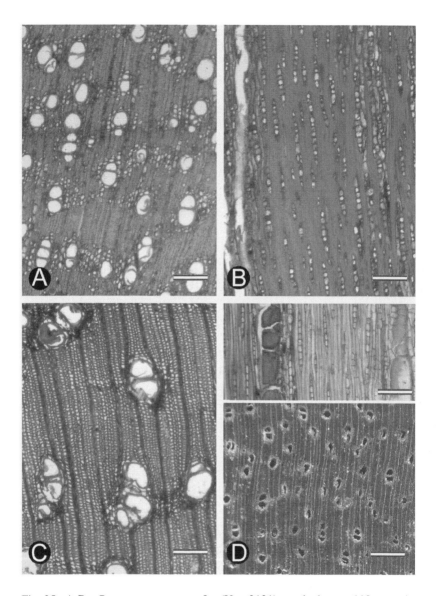

Fig. 35. A-B: *Conocarpus erectus* L. (Uw 2121), scale bar = 110 μm: A, transverse section; B, tangential section; C-D: *Laguncularia racemosa* (L.) C.F. Gaertn. (Uw 94a): C, transverse section, scale bar = 110 μm; D-above, tangential section, scale bar = 110 μm; D-below, transverse section, scale bar = 500 μm.

Parenchyma paratracheal, vasicentric to confluent and in bands of 3-5 cells wide, embedding the vessels; in marginal or in seemingly marginal bands; strands of (3-)5(-8) cells.

Ground tissue fibres thin- to thick-walled, sometimes very thick-walled; partly septate. Pits simple to minutely bordered, mainly on radial cell walls.

Material studied (Uw-numbers refer to the Utrecht Wood collection): **Conocarpus erectus** L.: USA-Florida: Kew Bot. Gardens s.n. (Uw 2121).

LAGUNCULARIA C.F. Gaertn.   – Fig. 35 C-D

Growth rings absent.

Vessels diffuse, solitary and in few radial multiples of clusters of 2-3(-6), sometimes intermingled with very narrow vessels; (9-)15(-23) per mm^2, tangential diameter (41-)120(-152) μm. Perforations simple with horizontal to oblique end walls. Intervessel pits alternate, round to polygonal, vestured, 6-9(-11) μm. Vessel-ray pits similar but infrequently tending to horizontal arrangement and elongate, up to 15 μm. Solid amorphic contents abundant.

Rays uniseriate, (9-)10(-12) per mm, composed of square and upright cells, with rare procumbent cells. Prismatic to elongate crystals solitary, in upright ray cells. Granular contents in many ray cells.

Parenchyma apotracheal scantily diffuse and paratracheal abundant, aliform to confluent, completely surrounding the vessels, sometimes in irregular 2-5 cells wide bands; strands of 3-7 cells.

Ground tissue fibres thin- to thick-walled, non-septate. Pits simple to minutely bordered, mainly on radial cell walls; rarely with amorphic contents.

Material studied (Uw-numbers refer to the Utrecht Wood collection): **Laguncularia racemosa** C.F. Gaertn.: Guyana: For. Dept. 5096 (Uw 854); Suriname: Stahel 94a (Uw 94a).

TERMINALIA L.   – Figs. 36 A-D; 37 A-D

Gowth rings absent to distinct.

Vessels diffuse, solitary and in radial multiples of 2-3 (10); varying from (3-)4(-5) per mm^2 in *T. dichotoma* to (9-)12(-16) in *T. catappa*, tangential

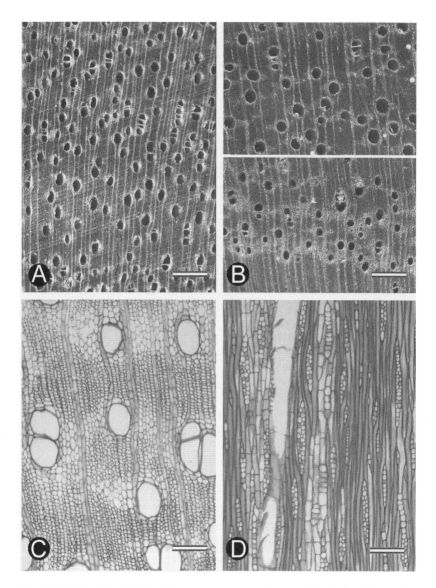

Fig. 36. A-D: *Terminalia*: A-B: transverse sections, scale bar = 500 µm: A, *T. amazonia* (J.F. Gmel.) Exell (Uw 6624); B, *T. catappa* L. (Uw 8403, above and Uw 8641, below); C-D: *T. catappa* L. (Uw 551), scale bar = 110 µm: C, transverse section; D, tangential section.

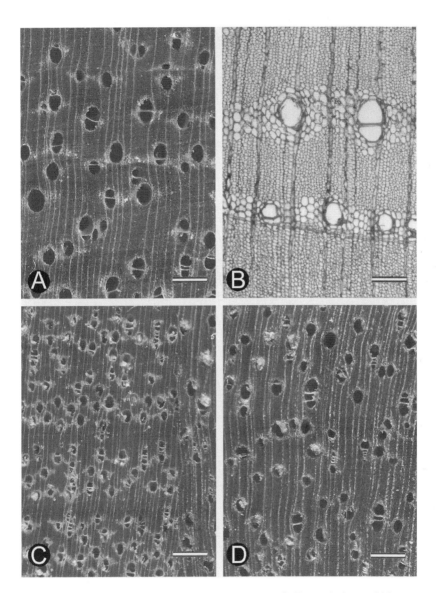

Fig. 37. A-D: *Terminalia*, transverse sections: A, C, D: scale bar = 500 µm; B, scale bar = 110 µm; A, *T. dichotoma* G. Mey. (Uw 363); B, *T. dichotoma* G. Mey. (Uw 3819); C, *T. guyanenesis* Eichler (Uw 1946); D, *T. quintalata* Maguire (Uw 16787).

174

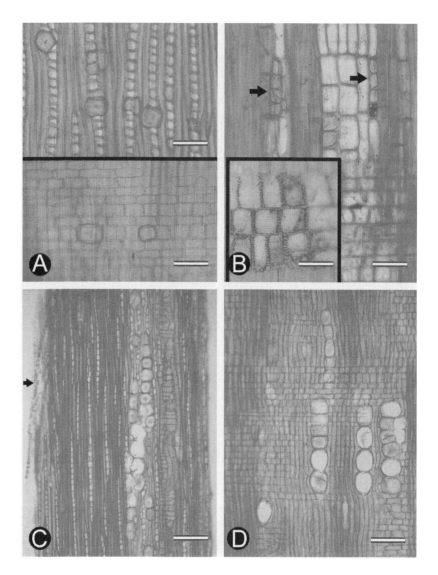

Fig. 38. A-D: Crystals in COMBRETACEAE: A, *Combretum laxum* Jacq. (Uw
22691), tangential section (above) and radial section (below), showing rhombic
crystals in enlarged ray cells, scale bar = 200 μm; B, *Terminalia guyanensis*
Eichler (Uw 1946), radial section, crystals in axial parenchyma, scale bar = 200
μm; B-insertion, *T. quintalata* Maguire (Uw 16787), radial section, ray cells with
disjunctive cell walls, scale bar = 200 μm; C-D: *Combretum fruticosum* (Loefl.)
Stuntz (Uw 30503), scale bar = 500 μm, with crystals in enlarged parenchyma
cells: C, tangential section, the arrow (left) points towards druses in bark tissue;
D, radial section.

diameter (55-)125(-180) μm. Perforations simple with horizontal to oblique end walls. Intervessel pits alternate, crowded, alternate, 9-11 μm, 6-9 μm and sometimes coalescent in *T. catappa*. Vessel-ray pits similar but half-bordered, tending towards horizontal arrangement. Vessel-parenchyma arrangement sometimes elongate, tending towards scalariform. Thin-walled tyloses as well as solid amorphic contents present in *T. catappa* and *T. dichotoma*.

Rays uniseriate, rarely with a 2(-3)-seriate part, 7-18 per mm, mainly composed of (weakly) procumbent cells with square and upright cells; disjunctive elements present in *T. quintalata*.

Parenchyma apotracheal diffuse and in narrow marginal bands 1-3 cells wide; paratracheal vasicentric, aliform-confluent in *T. amazonia*, aliform-confluent to banded in *T. catappa* p.p. and *T. dichotoma*; strands of 5-8 cells.

Crystals absent, or rarely present in ray- and/or axial parenchyma cells, large and rhomboidal, elongated rod- to styloidlike; druses reported for *T. catappa* in axial parenchyma.

Ground tissue fibres thin- to thick-walled; non-septate, but septate in *T. amazonica*. Gelatinous fibres frequent in *T. dichotoma*, infrequent in the other species. Pits simple to minutely bordered, 1-3 Ìm, mainly on radial cell walls.

N o t e : Our description, based on data for the 5 Guianan species studied here, suggests a relatively homogeneous wood structure. The wood of the large genus *Terminalia* as a whole, however, varies much more than can be concluded from the data given here. From van Vliet's work it appears that there is a huge variation in vessel features, ray composition and crystal forms. Part of this variation was intraspecific, or could be attributed to different habitats. See van Vliet (1979) for an elaborate description based on 43 species for which data were available.

M a t e r i a l s t u d i e d (Uw-numbers refer to the Utrecht Wood collection):

Terminalia amazonia (J.F. Gmel.) Exell: Suriname: Heyligers 441 (Uw 6624); Stahel 93 (Uw 93).

T. catappa L: Jamaica: Smiths. Inst. 6150 (Uw 8403); Suriname: Kramer & Hekking s.n. (Uw 8641); cult. Bot. Gard. U (Uw 482, Uw 551).

T. dichotoma G. Mey.: Guyana: Forest Dept. 2705 (Uw 855); Suriname: Stahel 363 (Uw 363); Lindeman 5522 (Uw 3819).

T. guyanensis Eichler: Suriname: Lanjouw & Lindeman 2853 (Uw 1946).

T. quintalata Maguire: Guyana: Maguire *et al.* 45805 (Uw 16787).

176

DICHAPETALACEAE

by

IMOGEN POOLE[25, 26, 27] & JIFKE KOEK-NOORMAN[27]

WOOD ANATOMY

The DICHAPETALACEAE are a family comprising 6 genera of small trees, shrubs or lianes found in lowland tropical regions around the world. In the Guianas 2 genera are represented, namely *Dichapetalum* and *Tapura*. The third genus, *Stephanopodium* Poepp. although found in tropical South America does not occur in the Guianas. *Dichapetalum* is represented by 3 species: the liana *D. pedunculatum* (DC.) Baill, *D. rugosum* (Vahl) Prance, which has either a liana or shrub habit, and the shrub *D. schulzii* Prance (Prance 1972). The wood of *Dichapetalum* has a coarse texture and is of no commercial value (Record & Hess 1943). *Tapura* is represented by 4 arboreal species, namely *T. amazonica* Poepp., *T. capitulifera* Baill. and *T. guianensis* Aubl., which can also be shrubby, and *T. singularis* Ducke. Although the wood of *Tapura* is relatively fine-textured with a straight grain, easy to cut and finishes smoothly it too is of no economic importance (Record & Hess 1943).

Terms are used in accordance with the defined descriptions according to the IAWA list of microscopic features for hardwood identification (IAWA Committee 1989).

GENERIC DESCRIPTIONS

DICHAPETALUM Thouars — Figs. 39, 40

Growth rings indistinct or absent.

[25] Department of Molecular Palaeontology, Faculty of Earth Sciences, Utrecht University, P.O. Box 80021, 3508 TA Utrecht, The Netherlands.
[26] Palaeoecology, Institute of Environmental Biology, Faculty of Science, Utrecht University, Laboratory of Palaeobotany and Palynology, Budapestlaan 4, 3584 CD Utrecht, The Netherlands.
[27] National Herbarium of the Netherlands, Utrecht University branch, Heidelberglaan 2, 3584 CS Utrecht, The Netherlands.

Acknowledgements: We are grateful to Lubbert Westra for providing the hand lens photographs of *Dichapetalum* and *Tapura*

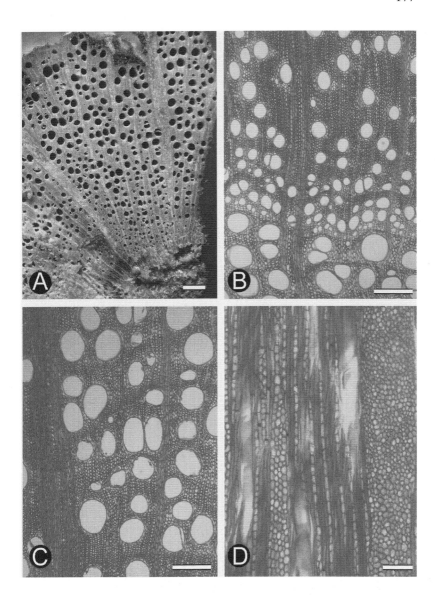

Fig. 39. A, B, D: *Dichapetalum pedunculatum* (DC.) Baill.: A, transverse section hand lens view, scale bar = 500 µm (Uw 36350); B, transverse section, scale bar = 250 µm (Uw 26350); D, tangential section, scale bar = 100 µm (Uw 26350); C: *Dichapetalum rugosum* (Vahl) Prance: C, transverse section, scale bar = 250 µm (Uw 12120).

178

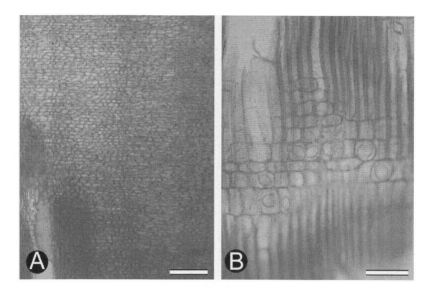

Fig. 40. A: *Dichapetalum rugosum* (Vahl) Prance: radial section, scale bar = 250 μm (Uw 12120); B: *Dichapetalum pedunculatum* (DC.) Baill.: radial section, scale bar = 100 μm (Uw 36350).

Vessels diffuse, mainly solitary, sometimes paired, 8-25 (10.5) per mm², circular or with flattened contact faces when paired, 49-284 (159) μm wide. Perforation plates simple. Intervessel pits and vessel-ray pits alternate with slit-like apertures; vessel-ray pits with distinct borders.
Rays uniseriate and multiseriate (up to 23), 3-8 per mm. Uniseriate rays composed of generally upright cells with some square to upright cells. Multiseriate rays heterogeneous composed predominantly of procumbent and square body ray cells and 1-10 (rarely more) rows of square to upright marginal cells. Multiseriate rays 32-483 (98) μm wide and from 188 μm to several millimetres high (excluding marginal wings). Sheath cells present. Prismatic crystals abundant in the ray cells.
Axial parenchyma mainly aliform (lozenge-aliform and winged aliform), but also diffuse and diffuse in aggregate, and 5-8 cells per strand. Ground tissue of non-septate fibres with thin-to-thick walls and abundant simple to minutely bordered pits on both the radial and tangential walls.

Note: Gelatinous fibres have been observed in young wood (Prance 1972).

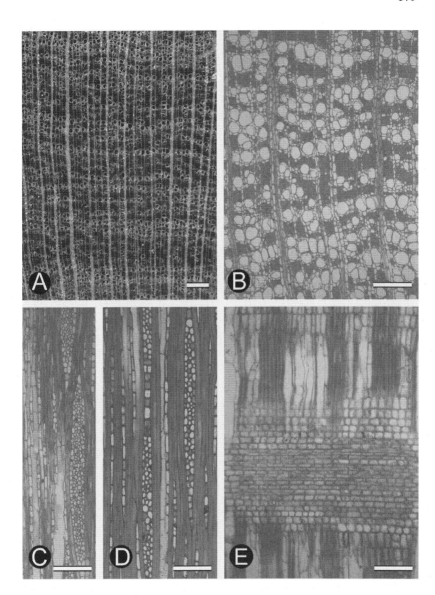

Fig. 41. A-C: *Tapura amazonica* Poepp.: A, transverse section hand lens view, scale bar = 500 μm (Uw 1726); B, transverse section, scale bar = 250 μm (Uw 1726); C, tangential section, scale bar = 250 μm (Uw 1726); D, E: *Tapura guianensis* Aubl.: D, tangential section, scale bar = 250 μm (Uw 5050); E, radial section, scale bar = 250 μm (Uw 5050).

180

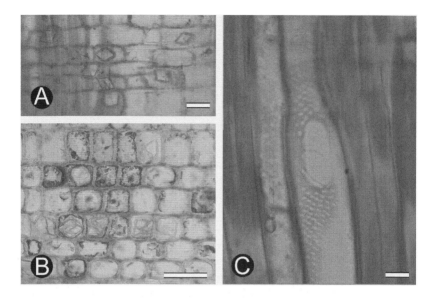

Fig. 42. A, C: *Tapura capitulifera* Baill.: A, radial section, scale bar = 25 μm
(Uw 133a); C, radial section, scale bar = 25 μm (Uw 133a); B: *Tapura
guianensis* Aubl.: B, radial section, scale bar = 100 μm (Uw 5050).

M a t e r i a l   s t u d i e d (Uw-numbers refer to the Utrecht Wood collection):
Dichapetalum pedunculatum (DC.) Baill.: Suriname: Lindeman &
Görts *et al.* 115 (Uw 26350); Guyana: van Andel 2164 (Uw 36350).
D. rugosum (Vahl) Prance: Suriname: van Donselaar 3488 (Uw 12120).

TAPURA Aubl.   – Figs. 41, 42

Growth rings indistinct or absent.
Vessels diffuse, mainly solitary (occasionally solitary in *T. capitulifera*),
sometimes paired (usually paired in *T. guianensis*) or in multiples (radial
groups in *T. capitulifera*) of up to 5 vessels; 31-59 (36) per mm², more
or less angular in outline, 41-100 μm (73) (up to 64 μm in *T. guianensis*
and 142 μm in *T. capitulifera*) wide. Perforation plates simple, rarely
with wide rims and/or with one bar. Intervessel pits and vessel-ray pits
alternate (rarely opposite in *T. capitulifera* and *T. guianensis*) with
minute apertures; vessel-ray pits with distinct borders.

Rays 1-3-seriate but sometimes multiseriate (3-5 cells), 7-11 per mm. Uniseriate rays composed of generally upright cells. Multiseriate rays heterogeneous composed predominantly of procumbent cells and 1-10, rarely more, rows of square to upright marginal cells. Multiseriate rays 32-92 (58) μm wide and from 488 up to 3000 (950) μm in height, sometimes composed. Prismatic crystals abundant in the ray cells.
Axial parenchyma apotracheal diffuse and forming short tangential bands, also paratracheal unilateral, vasicentric and aliform (winged aliform); 5-8, rarely more, cells per strand.
Ground tissue of non-septate fibres with thick walls and occasional simple to minutely bordered pits on the radial walls.

N o t e : Gelatinous fibres have been observed in young wood (Prance 1972); *T. capitulifera* exhibits fine strands running across some simple perforation plates.

M a t e r i a l   s t u d i e d (Uw-numbers refer to the Utrecht Wood collection):
Tapura amazonica Poepp.: Suriname: Lanjouw & Lindeman 2409 (Uw 1726).

T. capitulifera Baill.: Suriname: Stahel 133a (Uw 133a).

T. guianensis Aubl.: Suriname: Lindeman 3756 (Uw 2743); Schulz 7517 (Uw 5050).

LITERATURE ON WOOD AND TIMBER

Araujo, P.A. de M. & A. de Mattos Filho. 1974. Estutura das madeiras brasileiras de angispermas dicotiledôneas (VI). Cyrillaceae (Cyrilla antillana Michx.). Rodriguésia 27: 53-60.

Carlquist, S. 2001. Wood and stem anatomy of Rhabdodendraceae is consistent with placement in Caryophyllales sensu lato. IAWA J. 22: 171-181.

Détienne, P. & J. Jacquet. 1983. Atlas d'identification des bois de l'Amazonie et des régions voisines. Centre Technique Forestier Tropical, France, 348 pp + micrographs.

Edwards, K.S. & G.T. Prance. 2003. Four new species of Roupala (Proteaceae). Brittonia. 55: 61-68.

Heimsch, C.1942. Comparative anatomy of the secondary xylem in the Gruinales and Terebinthales of Wettstein with reference to taxonomic grouping. Lilloa 8: 83-198.

IAWA Committee. 1989. The IAWA list of microscopic features for hardwood identification. IAWA Bull. n.s. 10: 219-332.

InsideWood. 2004-onwards. Published on the Internet. http://insidewood/ lib.ncsu.edu/search.

Lens, F., S. Jansen, P. Caris, L. Serlet & E. Smets. 2005. Comparative wood anatomy of the primuloid clade (Ericales s.l.). Syst. Bot. 30: 163-183.

Mennega, A.M.W. 1966. Wood anatomy of the genus Euplassa and its relation to other Proteaceae of the Guianas and Brazil. Acta Bot. Neerl. 15: 117-129.

Prance, G.T. & V. Plana. 1998. The American Proteaceae. Austral. Syst. Bot. 11: 287-299.

Record, S.J. & R.W. Hess. 1936. Identification of woods with conspicuous rays. Trop. Woods 48: 1-30.

Record, S.J. & R.W. Hess. 1943. Timbers of the New World. Yale University Press, New Haven. 640 pp.

Schneider, E.L. & S. Carlquist. 2003. Unusual pit membrane remnants in perforation plates of Cyrillaceae. J. Torrey Bot. Soc. 130: 225-230.

Thomas, J.L. 1960. A monographic study of the Cyrillaceae. Contr. Gray Herb. 186: 3-114.

Vliet, G.L.C.M. van, 1976. Radial vessels in rays. IAWA Bull. 3: 35-37.

Vliet, G.L.C.M. van, 1978. The vestured pits of Combretaceae and allied families. Acta Bot. Neerl. 27: 273-285.

Vliet, G.L.C.M. van, 1979. Wood anatomy of the Combretaceae. Blumea 25: 141-223.

TAXONOMIC CHANGES
NEW TYPIFICATIONS

Cyrillaceae

Lectotypification:
Cyrilla breviflora N.E. Br.

Combretaceae

Lectotypification:
Combretum aubletii DC.

NUMERICAL LIST OF ACCEPTED TAXA

Cyrillaceae

1. Cyrilla Garden ex L.
   1-1.   C. racemiflora L.

Theophrastaceae

1. Clavija Ruiz & Pav.
   1-1.   C. imatacae B. Ståhl
   1-2a.   C. lancifolia Desf. subsp. lancifolia
   1-2b.   C. lancifolia Desf. subsp. chermontiana (Standl.) B. Ståhl
   1-3.   C. macrophylla (Link ex Roem. & Schult.) Miq.

Rhabdodendraceae

1. Rhabdodendron Gilg & Pilg.
   1-1.   R. amazonicum (Spruce ex Benth.) Huber

Proteaceae

1. Euplassa Salisb. ex Knight
   1-1.   E. glaziovii (Mez) Steyerm.
   1-2.   E. pinnata (Lam.) I.M. Johnst.

2. Panopsis Salisb. ex Knight
   2-1.   P. rubescens (Pohl) Rusby
   2-2.   P. sessilifolia (Rich.) Sandwith

3. Roupala Aubl.
   3-1.   R. montana Aubl.
   3-2.   R. nitida Rudge
   3-3.   R. obtusata Klotzsch
   3-4.   R. soropana Steyerm.
   3-5.   R. suaveolens Klotzsch

Combretaceae

1. Buchenavia Eichler
   - 1-1.　　B. congesta Ducke
   - 1-2.　　B. fanshawei Exell & Maguire
   - 1-3.　　B. grandis Ducke
   - 1-4.　　B. guianensis (Aubl.) Alwan & Stace
   - 1-5.　　B. macrophylla Eichler
   - 1-6.　　B. megalophylla Van Heurck & Müll. Arg.
   - 1-7.　　B. nitidissima (Rich.) Alwan & Stace
   - 1-8.　　B. ochroprumna Eichler
   - 1-9.　　B. pallidovirens Cuatrec.
   - 1-10.　 B. parvifolia Ducke
   - 1-11.　 B. reticulata Eichler
   - 1-12.　 B. tetraphylla (Aubl.) R.A. Howard
   - 1-13.　 B. viridiflora Ducke

2. Combretum Loefl.
   - 2-1.　　C. cacoucia Exell
   - 2-2.　　C. fruticosum (Loefl.) Stuntz
   - 2-3.　　C. fusiforme Gleason
   - 2-4.　　C. laxum Jacq.
   - 2-5.　　C. paraguariense (Eichler) Stace
   - 2-6.　　C. pyramidatum Ham.
   - 2-7.　　C. rohrii Exell
   - 2-8.　　C. rotundifolium Rich.
   - 2-9.　　C. spinosum Bonpl.

3. Conocarpus L.
   - 3-1.　　C. erectus L.

4. Laguncularia C.F. Gaertn.
   - 4-1.　　L. racemosa (L.) C.F. Gaertn.

5. Terminalia L.
   - 5-1.　　T. amazonia (J.F. Gmel.) Exell
   - 5-2.　　T. catappa L.
   - 5-3.　　T. dichotoma G. Mey.
   - 5-4.　　T. guyanensis Eichler
   - 5-5.　　T. lucida Hoffmanns. ex Mart. & Zucc.
   - 5-6.　　T. oblonga (Ruiz & Pav.) Steud.
   - 5-7.　　T. quintalata Maguire

Dichapetalaceae

1. Dichapetalum Thouars
   1-1.   D. pedunculatum (DC.) Baill.
   1-2.   D. rugosum (Vahl) Prance
   1-3.   D. schulzii Prance

2. Tapura Aubl.
   2-1.   T. amazonica Poepp.
   2-2.   T. capitulifera Baill.
   2-3.   T. guianensis Aubl.
   2-4.   T. singularis Ducke

Limnocharitaceae

1. Hydrocleys Rich.
   1-1.   H. nymphoides (Willd.) Buchenau

Alismataceae

1. Echinodorus Rich. ex Engelm.
   1-1.   E. bolivianus (Rusby) Holm-Niels.
   1-2.   E. grisebachii Small
   1-3.   E. horizontalis Rataj
   1-4.   E. macrophyllus (Kunth) Micheli subsp. scaber (Rataj) R.R. Haynes & Holm-Niels.
   1-5.   E. paniculatus Micheli
   1-6.   E. reticulatus R.R. Haynes & Holm-Niels.
   1-7a.   E. subalatus (Mart.) Griseb. subsp. subalatus
   1-7b.   E. subalatus (Mart.) Griseb. subsp. andrieuxii (Hook. & Arn.) R.R. Haynes & Holm-Niels.
   1-8.   E. tenellus (Mart.) Buchenau

2. Sagittaria L.
   2-1.   S. guayanensis Kunth subsp. guayanensis
   2-2.   S. lancifolia L. subsp. lancifolia
   2-3.   S. rhombifolia Cham.

COLLECTIONS STUDIED
(Collection numbers in **bold** refer to types)

Cyrillaceae

GUYANA

Abbersetts, N.J., 19, 35 (1-1)
Anderson, C.W., 543 (1-1)
Bio, U.G., 106 (1-1)
Cooper, A., 141 (1-1)
Cowan, R.S. & T.R. Soderstrom, 1956, 2229 (1-1)
Cruz, J.S. de la, 1822, 2080, 3467 (1-1)
Davis, D.H., 98, 182, 634 (1-1)
Fanshawe, D.B., 5242 (= F2506) (1-1)
Forest Dept. British Guiana, 933, 2867, 7804, 7924 (1-1)
Gillespie, L.J., *et al.*, 917, 1342, 2604, 2671, 2942 (1-1)
Gleason, H.A., 486 (1-1)
Graham, V., 113 (1-1)
Hahn, W., *et al.*, 4046, 5359, 5521 (1-1)
Henkel, T.W., *et al.*, 2459, 2497 (1-1)
Hoffman, B., *et al.*, 1610, 1870 (1-1)
Hohenkerk, L.S., 250A (1-1)
Im Thurn, E.F., (set. B) 35, **334** (1-1)
Jenman, G.S., s.n. (sep 1888), 288, 1044, 1861, 4350 (1-1)
Kelloff, C.L., *et al.*, 875 (1-1)
Kvist, L.P., *et al.*, 79 (1-1)
Liesner, R., 23327 (1-1)
Maas, P.J.M., *et al.*, 4344, 5701, 5752 (1-1)
Maguire, B., *et al.*, 23106, 32155, 46025A (1-1)
McDowell, T., *et al.*, 2831, 3045, 4590, 4789 (1-1)

Mori, S., *et al.*, 8044, 8290 (1-1)
Mutchnick, P., *et al.*, 182 (1-1)
Persaud, R., 133 (1-1)
Pinkus, A.S., 190 (1-1)
Pipoly, J.J., *et al.*, 7392, 7842, 8423, 9516, 9540, 9793, 9992, 10755, 11402, 11538 (1-1)
Prance, G.T., 16565 (1-1)
Quelch, M.J. & F. McConnell, 88, 193, 203, 318, 638 (1-1)
Renz, J., 14276-A, 14282 (1-1)
Schomburgk, Ro., ser. I, 219, ser. II, 596 (1-1)
Tate, G.H.H., 231, 387A, 401 (1-1)
Tillett, S.S., *et al.*, 45102, 45730 (1-1)
Tiwari, S. & A. Mengharini, 788 (1-1)
Warrington, J.F., *et al.*, (K.E.R. 130) (1-1)

SURINAME

BW (Boschwezen), 946, 3429 (1-1)
Maas, P.J.M., LBB 10868 (1-1)

FRENCH GUIANA

Oldeman, R.A.A., B-1474 (1-1)

Theophrastaceae

GUYANA

Forest Dept. British Guiana (FD), 202 (1-2b); 219 (1-3); F2058 (1-1)

188

Goodland, R.J.A. *et al.*, 462
(1-2b)
Jansen-Jacobs, M.J. *et al.*, 251
(1-2b); 356, 2638 (1-3); 4458
(1-2b); 4845 (1-3)
Maas, P.J.M. *et al.*, 3786 (1-2b)
Sandwith, N.Y., 635 (1-1)
Schomburgk, Ro., ser. II, 419 =
Ri. 659, ser. I, 750 (1-3)
Schomburgk, Ri., 659 = Ro.
ser.II, 419 (1-3)
Smith, A.C., 2365, 3172 (1-3);
3218 (1-2b)

SURINAME

Boon, H., 1013, 1138 (1-2b)
BW, 101, 500 (1-2a); 2173, 3668
(1-2b); 3857, 5290 (1-2a);
6168 (1-2b); 6667 (1-2a)
Daniëls, A.G.H. & F.P. Jonker
991, 1023 (1-2b)
Florschütz, P.A. & J., 2265 (1-2b)
Florschütz, P.A. & P.J.M. Maas,
2414, 2475 (1-2b)
Hekking, W.H.A., 987 (1-2b)
Irwin, H. *et al.*, 54617, 55114,
55582, 55892, 55898 (1-2b)
Kappler, A., s.n. (1-2a)
Kegel, H.A.H., 1007 (1-2b)
Lanjouw, J., 807 (1-2b)
Lindeman, J.C., 4463 (1-2a);
5003, 6308, 6693, 6813 (1-2b)
Pulle, A., 514 (1-2b)
Schulz, J.P., 7713, 7963, 8016
(1-2b)
Tresling, J. 253 (1-2b)
Tulleken, J.E., 329 (1-2b)
Went, F.C., 104 (1-2b)
Wullschlägel, H., 1050 (1-2b)

FRENCH GUIANA

Aubréville, A., 152 (1-2b)
Cremers, G.A., 4680, 6464 (1-2b)
Granville, J.J. de, 657, 1014, 2149,
2150, 2211, 3707, 4990 (1-2a);
5370 (1-2b); 5581 (1-2a)
Grenand, P., 1041 (1-2b)
Hallé, F., 1034 (1-2a)
Leprieur, F.R., s.n. (1-2a)
Mori, S.A. *et al.*, 17928, 20731,
22838 (1-2b)
Oldeman, R.A.A., 806, 2905,
2910, 3043 (1-2a); 3240,
3992, 4035, 4087 (1-2b)
Poiteau, P.A., **s.n.** (1-2a)
Sagot, P., 1252 (1-2a)
Sastre, C., 1409 (1-2b)

Rhabdodendraceae

GUYANA

Chanderbali, A., 53 (1-1)
Clarke, D., 706, 3323 (1-1)
Ek, R.C., 572)1-1)
Fanshawe, D.B., 138, 1314 (1-1)
Forest Dept. British Guiana,
2747 (1-1)
Gleason, H.A., **211** (1-1)
Hohenkerk, L.S., 450, 450A (1-1)
Maguire, B. & D.B. Fanshawe,
23558, 32141 (1-1)
McDowell, T., 3251 (1-1)
Pipoly, J.J., 8584, 8947, 8963
(1-1)
Raes, N., *et al.*, 25 (1-1)
Sandwith, N.Y., 1237 (1-1)
Steege, H. ter, *et al.*, 399 (1-1)
Stoffers, A.L. & A.R.A. Görts-
van Rijn *et al.*, 60, 150 (1-1)
Tutin, T.G., 243 (1-1)

SURINAME

Acevedo-Rodriguez, P., 6067 (1-1)
LBB, 10694 (1-1)
Lindeman, J.C. & A.L. Stoffers et al., 157, 785 (1-1)
Mori, S.A. & A. Bolten, 8542 (1-1)
Schulz, J.P., 8322a (1-1)
Teunissen, P.A., in LBB 15941 (1-1)
Wessels Boer, J.G., 334 (1-1)

FRENCH GUIANA

Feuillet, C., 1209 (1-1)
Martin, J., **s.n.** (1-1)
Sastre, C., 4433 (1-1)
Service Forestier, 7406 (1-1)

Proteaceae

GUYANA

Appun, C.F., 1203 (1-1)
Clarke, D., 3304 (2-1); 6946 (3-5)
Cook, C.D.K., 36 (3-1)
Cruz, J.S. de la, 2738, 3075 (2-2)
Forest Dept. British Guiana, 2254 (2-1); 2814 (3-5); 3040, 5288, 5938 (2-2)
Goodland, R., *et al.*, 406, 694, 875 (3-1)
Hahn, W., 5733 (3-1)
Jansen-Jacobs, M.J., *et al.*, 105, 524, 1057 (3-1); 1628 (3-5); 1737 (2-1)
Jenman, G.S., 979 (2-1); 3918, 5328, 7594 (2-2)
Knapp, S., *et al.*, 2783 (3-1)
Lanjouw, J. & J. van Donselaar, *et al.*, 839 (2-1)

Maas, P.J.M., *et al.*, 3813 (2-1); 3845, 5635 (3-1); 7630 (2-1)
Maguire, B., *et al.*, 23372, 32434 (2-2)
Mutchnick, P., 440 (2-2)
Persaud, A.C., 71 (2-2)
Pipoly, J.J., 10378 (3-1)
Sandwith, N.Y., 450 (2-1); 472, 483 (2-2)
Schomburgk, Ri. or Ro., s.n. (2-2)
Schomburgk, Ri., **215** (3-3); **855** (3-5); **1045** (3-5); 1046B (3-5)
Schomburgk, Ro., ser. I, 69 (2-1); ser. II, **544** (3-5); ser. II, 634 (2-2)
Smith, A.C., 2208 (3-1); 2568, 2674 (2-1); 3360 (3-1)
Tillett, S.S., *et al.*, 45510 (2-2)

SURINAME

Anderson, A., s.n. (2-2)
Evans, R., *et al.,* 2401 (2-2)
Geyskes, D.C., 17 (2-1)
Hulk, J.F., **333** (3-5)
Irwin, H.S., *et al.*, 55356 (2-2); 55692 (3-1); 55878 (2-2)
Kappler, A., 1999, 2031 (2-2)
Lanjouw, J., 1256 (3-1)
Lanjouw, J. & J.C. Lindeman, 2759 (1-2); 2872 (2-1)
Lindeman, J.C., 6413 (2-2)
Lindeman, J.C. & R.S. Cowan, 7007 (1-2)
Lindeman, J.C. & A.L. Stoffers, *et al.*, 487 (3-1)
Mennega, A.M.W., 464 (2-2)
Mori, S.A., *et al.*, 15168 (1-2)
Oldenburger, F.H.F., *et al.*, 145, 298 (3-1); 1072 (2-1)
Rombouts, H.E., 721 (2-1)
Schulz, J.P., LBB **10307** (3-4)
Stahel, G., 249 (3-1); 291 (2-2)

190

Teunissen, P.A., *et al.*, LBB
11834 (3-1); LBB 15447 (2-2)
Tresling, J.H.A.T., 357 (2-2)
Troon, F. van, LBB 16321 (2-2)
Versteeg, G.M., 233 (2-2)
Wessels Boer, J.G., 992 (2-1)
Wullschlägel, H.R., 1677 (2-2)

FRENCH GUIANA

Aublet, J.B.C.F., **s.n.** (3-1)
BAFOG, 1208, 1266 (1-2); 1322
(2-2)
Benoist, R., 49 (2-2)
Cremers, G., 7830 (3-1)
Granville, J.J. de, 481(3-2)
Grenand, P., 501 (2-2); 626 (3-2)
Herb. Mus. Paris, 421 (2-2)
Leblond, J.B., s.n. (2-1); **s.n.**
(**224**) (2-2)
Leprieur, F.R., **s.n.** (2-2)
Martin, J., **s.n.** (2-2); **s.n.** (3-1);
s.n., 52 (3-2)
Mélinon, E., s.n. (3-1); 161, 179
(2-2)
Moretti, C., 822 (2-2)
Oldeman, R.A.A., T-65, T-959,
B-1174, B-2049, 2432 (2-2);
2692, 3056, B-3124 (3-2)
Perrottet, G.S., s.n. (3-1)
Prévost, M.F., *et al.*, 1043 (3-2)
Richard, L.C., **s.n.** (1-2)
Rohr, J.P.B. von, **s.n.** (3-1)
Sabatier, D., *et al.*, 1034, 2114,
4011 (2-2)
Sagot, P.A., s.n. (2-2)

Combretaceae

GUYANA

Acevedo-Rodríguez, P., *et al.*,
3279 (2-8); 3302 (2-4); 3319
(2-8); 5812 (2-4)

Andel, T. van, *et al.*, 1160 (2-8);
1280 (5-3); 1734 (1-1); 3008
(2-1)
Anderson, C.W., 26 (5-3); 79 (2-
1); 210, 390 (5-3); 749, s.n.
(1-2)
Appun, C.F., 288 (2-1)
Archer, W.A., 2384 (2-4); 2373
(2-1); 2404 (2-4)
Atkinson, D.J., 555 (5-1)
Bancroft, C.K., *et al.*, s.n. (2-1)
Chanderbali, A., 17 (2-8)
Clarke, D., *et al.*, 271 (1-6); 357,
1378 (2-8); 1380 (1-6); 1658,
1771 (2-2); 2230 (1-2); 2757
(1-6); 2232, 3013 (2-4); 3303,
3463 (2-6); 3476 (2-8); 3531
(2-6); 3534 (1-2); 3787 (2-6);
3807 (1-2); 3873, 3924 (2-8);
3926, 4756 (2-6); 4771 (1-2);
4778 (2-2); 5060 (1-2); 5074
(2-4); 6191 (5-3); 6437 (2-4);
6497 (2-2); 6661 (5-3); 7054,
7056 (2-6); 7731, 7737, 8009,
8712, 8875 (2-4)
Cook, C.D.K., 247, 248 (2-2)
Cruz, J.S. de la, 1094 (2-4); 1114
(2-1); 1141 (2-4); 1333, 1440
(5-3); 1488, 1509 (2-1); 1863
(1-2); 1947(2-4); 2365, 2428,
2680 (2-1); 2979 (5-3); 3074
(1-2); 3307, 3425, **3566**,
3585, 3713 (2-4); **4104** (2-3);
4129 (2-4); 4156 (5-3); 4258
(2-4); 4578 (5-3); 4580 (2-4)
Davis, T.A.W., 1062 (5-1)
Dirven, J.G.P., LP194 (2-8)
Forest Dept. British Guiana,
2152 = D-161, 2252 = D-261
(5-1); 2378 (1-2); 2705 = F-
106 (5-3); 3513 = F-777 (5-7);
3572 = F-836 (2-1); 3580 = F-
844 (1-2); 3581 p.p. = F-845
p.p. (1-6); 3775 = field no.

192

King, K.F.S., 6399 (3-1)
Kelloff, C.L., *et al.*, 1058 (5-7)
Knapp, S., *et al.*, 2780 (2-2)
Lanjouw, J., *et al.*, 874 (2-8)
Leechman, A., s.n. (3-1)
Little, E.L., 16949 (2-1)
Maas, P.J.M., *et al.*, 3596, 3619,
    4059 (2-4); 4095 (2-2); 5570
    (2-1)
Maguire, B., *et al.*, 23313 (5-7);
    23355 (2-4); 23459 (5-7);
    23499 (1-2); **23551**, 32640,
    32640, 40656 (5-7)
Martyn, E.B., 247 (5-1)
McDowell, T., *et al.*, 1829, 1833
    (2-2); 2071 (2-4); 2135 (2-4);
    2143, 2186 (2-2); 2302 (2-8);
    2846 (5-7); 3329 (2-8); 3399
    (2-6)
Meyer, G., **113** (5-3)
Meyer, P.E., 45 (2-1)
Mori, S.A., *et al.*, 8153 (5-3);
    24347 (1-6)
Mutchnick, P., *et al.*, 373 (3-1);
    440 (1-6); 458 (2-4); 506, 512
    (2-8); 565 (2-4); 770 (1-6);
    835 (2-4); 988 (2-8); 1015 (2-
    4); 1269 (2-8); 1279 (1-6);
    1295 (2-4); 1359 (2-8)
Omawale, *et al.*, 42 (5-2)
Parker, C.S., s.n. (1-6); s.n. (2-4);
    s.n. (5-2)
Pennington, T.D., *et al.*, 345 (2-
    8); 410 (5-1)
Persaud, A.C., 35, 56 (2-6); 187
    (5-3); 223 (1-2); 389 (5-2)
Pipoly, J.J., *et al.*, 7438, 7459 (5-
    1); 7795 (5-7); 11247 (3-1);
    11717 (2-1); 11751 (3-1);
    11777 (5-1)
Polak, M., *et al.*, 119 (5-3); 302
    (5-1); 316 (1-9); 521 (1-2);
    576 (1-2)
Reinders, M.A., *et al.*, 159 (2-4)

Rudge, E., s.n. (1-12)
Sandwith, N.Y., 104 (2-4); 475
    (5-3); 619 (2-1); 1219 (1-2)
Schomburgk, Ro., ser.I, 272 (2-
    1); **ser. I, 87 pro parte** (2-2);
    ser. I, 87 pro parte (2-8); ser.
    II, 872 (2-4)
Singh, 1 (6389) (3-1)
Smith, A.C., 2118 (1-6); 2833
    (1-10); 3033 (1-2)
Stergios, B., *et al.*, 5272, 5294
    (2-9) (VEN)
Stoffers, A.L., *et al.*, 9 (2-4); 348
    (2-2)
Talbot, H.F., s.n. (1-12); s.n.
    (2-4)
Taylor, 85 (1-6)
Thurn, E.F. im, s.n. (2-1); s.n.
    (2-4)
Tillett, S.S., *et al.*, 45805 (5-7)
Tiwahri, S., *et al.*, 376 (2-1); 961
    (5-3)
Watson, A.G., 31 (2-1); 32 (2-8);
    39, 1939 (2-4)

SURINAME

Acevedo-Rodríguez, P., *et al.*,
    5812 (2-4); 5954 (1-12); 6006
    (5-1); 6160 (1-6)
BBS (Bosbeheer Suriname), 91
    (1-12); 158 (5-3)
BW (Boschwezen), 374, 1488,
    2026 (5-1); 2035, 3343 (2-1);
    1312, 3962, 4085, 4661 (1-
    12); 5471 (2-8); 5707 (2-4);
    6000, 6261 (1-12)
Berthoud-Coulon, M., 576 (4-1)
Boerboom, J.H.A., 9140 (5-1)
Boon, H.A., 1257 (2-1)
Brownsbay, 6594 (5-1)
Daniëls, A.G.H. & F.P. Jonker,
    1326 (2-4)

194

Tulleken, J.E., 409 (2-1)
Versteeg, G.M., 540 (2-1); 553 (5-2); 647 (5-3); 707 (2-8)
Vreden, C., LBB, 15774 (5-1); 16204 (5-5)
Went, F.A.F.C., 123 (2-6); 126 (2-1); 484 (5-2)
Wessels Boer, J.G., 232 (2-8); 444 (2-4)
Wullschlägel, H.R., 152, s.n. (2-1); 878 (5-3); 1445 (5-5)

FRENCH GUIANA

Aublet, J.B.C.F., 281 (5-3); **s.n.** (1-4); **s.n.** (2-1); s.n. (2-8); **s.n.** (5-3)
Aubréville, A., 222 (5-1); 333 (2-8)
Barrabé, L., *et al.* 174 (4-1)
Barrier, S., 4901 (1-7); 5059, 5080 (5-1)
Béna, P., in BAFOG 1313 (1-8)
Benoist, R., 2 (1-12); 780 (3-1); 1256 (5-2)
Billiet, F., *et al.*, 1012 (4-1); 1260, 1897 (2-1); 4496 (2-4); 7492 (5-1)
Black, G.A., *et al.*, 54-17593 (4-1)
Bois, 7585 (5-1)
Boom, B.M., *et al.*, 2118 (5-3); 2134, 2237 (5-4); 2507 (5-3)
Bordenave, B., 375 (2-8); 784 (2-4); 816 (1-12); 826 (2-4); 850 (5-3); 855B (1-7); 1014 (1-12)
Broadway, W.E., 536 (1-12); 3110 (5-3)
Cremers, G., *et al.*, 7192 (2-4); 7198 (2-8); 8401 (4-1); 9332 (5-5); 13732 (2-8); 13739 (2-6)
Desfontaines, R.L., s.n. (5-3)

Desvaux, A.N., **s.n.** (2-6)
Feuillet, C., 1154 (5-1); 2366 (2-1)
Fleury, M., 356 (5-1); 446, 1121, 1154 (2-8)
Foresta, H. de, 476 (5-5)
Gabriel, A., 18002 (5-2)
Garnier, F.A., 109 (5-1); 195 (1-12)
Geay, F.,1903, 1904, 1905 (2-1)
Girod *et al.*, 3237 (2-1)
Granville, J.J. de, *et al.*, 77 (2-8); T-1036 (2-8); 3948 (5-1); B-4691 (5-3); B-4792 (1-6); B-5267 (5-3); 5768 (5-1); 6852 (2-1); 6854, 7153 (2-4); 8176 (2-8); 9296 (1-12); 9599 (2-8); 9600 (2-4); 9614 (1-5); 9671 (2-4); 10958 (5-4); 11889 (2-4); 12581 (2-8); 15272 (1-12)
Grenand, P., 78 (1-13); 1293 (5-1); 1303 (5-6); 1306, 1401 (5-1); 1734 (1-12); 3014 (5-1); 3065 (5-4)
Hallé, F., 538 (2-8); 539 (2-1); 545, 600 (2-4); 2772 (5-3); 4013 (2-4)
Hoff, M., 5012 (5-5); 5087 (5-2); 5090 (5-5); 5105 (4-1); 5982 (2-8)
Irwin, H.S., *et al.*, 48328 (1-12)
Jacquemin, H., 2844 (4-1)
Leblond, J.B., 83 (2-1); 116 (2-4); **117** (2-8); 403, 437 (1-12); **s.n.** (1-7); **s.n.** (2-4); s.n. (2-7)
Le Goff, A., 63(4-1)
Lemée, A.M.V., 1901 (5-5); s.n. (2-1)
Lemoine, S.V., 7774 (2-8); 7883 (2-4); 7897 (1-5); 7913 (2-4)
Lemoult, E., 9 = Service Forestier, 4172 (2-1)

Leprieur, F.R.M., s.n. (1-12); s.n. (2-1); s.n. (3-1); s.n. (4-1); s.n. (5-5)

Lescure, J.P., 122 (1-7); 272, 689 (5-2)

Loubry, D., 1215, 1934 (1-12); 1949 (2-6)

Martin, J., 110 (2-1); 169 (2-4); 197, 450 (2-1); s.n. (1-12); s.n. (2-1); s.n. (2-4); s.n. (2-6); s.n. (5-2)

Mélinon, E.M., 45, 69 (2-1); 134 (5-3); 206 (4-1); 291 (2-4); 296, 308, 381, 444, s.n. (2-1); s.n. (5-3)

Menten, s.n. (5-2)

Moretti, C., 634 (2-4); 824 (1-5)

Mori, S.A., et al., 8794 (1-13); 8948 (2-4); 15095 (1-7); 15121, 15223 (5-4); 15300 (1-10); 15343 (1-7); 18625 (2-7); 18653 (5-4); 20763, 20941 (2-4); 22128 (2-1); 22129, 22140, 22642 (2-4); 23905 (5-4); 23485 (1-3); 23905 (5-4); 25074 (5-1)

Nolland, L.A., s.n. (5-3)

Oldeman, R.A.A., 97 (2-4); T-269 (2-4); 282 (5-3); T-324 (2-4); B-673 (2-7); B-762 (2-1); T-877 (2-4); 972 (5-3); B-1149 (2-4); 1225 (2-8); 1323 (2-8); 1371 (2-4); B-1377, B-1400 (2-4); 1403 (2-1); 1406 (2-8); B-1421, 1435 (2-4); B-1531 (2-5); 1547 (2-4); B-1783 (5-3); 1791 (2-5); 1850 (2-4); B-1926 (5-3); B-2136 (2-4); B-2157 (5-3); 2252 (5-1); B-2308 (1-13); 2451 (5-3); B-2472 (1-10); 2473, 2516 (1-7); B-2552 (2-8); 2637 (2-5); B-2648 (2-4); 2652 (2-5); 2661 (2-4); B-2809 (2-1); 2825 (5-1); 2981 (5-3);

B-3151 (2-4); 3190 (5-4); B-3403, B-3611, B-3682 (2-4)

Patris, J.B., s.n. (2-4); **s.n.** (5-1)

Perrottet, G.S., s.n. (1-12); s.n. (2-7); **s.n.** (5-1)

Petrov, I., 203, 208 (5-3); 234 (5-1)

Poncy, O., et al., 1874 (1-4)

Poiteau, P.A., s.n. (2-1); s.n. (2-8); s.n. (5-3); **s.n.** (5-4)

Prévost, M.F., et al., 329 (2-1); 977 (5-3); 1221 (5-5); 1312, 1313, 1359 (2-8); 1781 (2-4); 1975 (5-1); 2162 (5-2); 2307 (5-3); 2824 (5-1); 3564, 3648 (2-4); 3703 (2-1); 4325 (1-13); 4357 (1-3); 4550 (3-1); 4895 (5-5)

Raynal-Roques, A., et al., 24780 (2-4); 24781 (2-3)

Richard, L.C.M., s.n. (1-6); s.n. (2-1); s.n. (2-3); s.n. (2-4); s.n. (2-7); s.n. (5-2); s.n. (5-5)

Riera, B.J.J., 496 (5-1)

Rohr, J.P.B. von, **149** (2-7); 158 (5-5); 435 (2-4)

Sabatier, D., et al., 24 (2-8); 575 (5-4); 997 (2-6); 1019 (2-4); 1134 (1-3); 1151 (1-7); 1679, 1681 (1-7); 2179 (5-1); 2259 (4-1); 2264 (1-12); 2283 (1-4); 2298 (5-4); 2299 (1-7); 2300 (1-10); 2309 (5-4); 2582 (1-4); 2627 (5-3); 2692 (1-12); 2837 (5-3); 2862, 3355, 3515 (5-1); 3679 (1-12); 4188 (5-1); 4677 (1-12)

Sagot, P.A., 1 (2-4); 240 (4-1); 241 (2-4); 242, 254 (2-1); 255, 944 (2-4); 794 (5-2); 795 (5-3); 944 (2-4); 1007 (5-5); s.n. (2-1); s.n. (5-3)

Sastre, C., et al., 4117 (2-8); 5761 (5-5); 6364 (2-1); 8129 (2-8); 8198 (2-6); 8199 (2-8)

Sauvain, M., 3 (5-5)
Service Eaux et Forêts, 4050
(2-6)
Service Forestier, 4047 (5-3);
4172 (2-1); 4303 (2-4); 7774,
7832 (2-8); 7883 (2-4); 7897
(1-5); 7913 (2-4); s.n. (5-3)
Skog, L.E., *et al.*, 7456 (2-4)
Vahl, M. (herb.), s.n. (2-1)
Wachenheim, H., 52 (2-1)
Weitzman, A.L., *et al.*, 247 (2-1)

Dichapetalaceae

GUYANA

Abraham, A.A., 249 (2-3)
Acevedo Rodriguez, P., *et al.*,
3423 (2-3)
Aitkin, J.B., 40 (1-1)
Archer, W.A., 2397, 2468 (1-1)
Beckelt, J.E. & P. Kartright,
8534 (1-1)
Boom, B.M. & G.J. Samuels,
8907 (2-3)
Boyan, J., 52 = FD 7736 (2-2)
Clarke, D., 522, 591 (2-3); 751
(2-2); 1269, 1468, 2478,
2724, 2717, 2882, 3260, 3415,
3613, 3690 (2-3); 3733 (1-2);
3975, 4197 (2-3); 4440 (2-2);
4635, 5528, 6049 (1-1); 6081,
6348, 7621 (2-3)
Cruz, J.S. de la, 1108, 1193,
1596, 1895 (1-1); 1917 (2-3);
2625, 3095 (1-1); 3304 (2-3);
3543, 3688, 3826, 3905, 4041,
4337, 4517 (1-1)
Davis, T.A.W., 85, 557 (2-3)
Ehringhaus, C., *et al.*, 69 (2-3)
Ek, R.C., *et al.*, 597, 740, 1004
(2-3)

Fanshawe, D.B., 143, 182 (2-3);
205, 230 (1-1); 585 (2-3);
1079 (1-1); 1322, 1565, 2744,
2752 (2-3)
Forest Dept. British Guiana (FD)
(see also Boyan), 935 (2-3);
1078 (1-1); 2715, 2752 (2-3)
Gillespie, L.J., 1584 (2-3)
Gleason, H.A., 205 (1-1)
Hahn, W., 3658, 5190, 5640 (2-
3); 5661 (1-1); 5834 (2-2)
Henkel, T.W., *et al.*, 546, 1833,
2043 2087, 4660 (2-3)
Hitchcock, A.S., 17084 (2-3)
Hoffman, B., *et al.*, 472, 1168,
2569, 3505, 3526, 4544 (2-3)
Im Thurn, E.F., s.n. (2-3)
Irwin, H.S., 198 (1-1)
Jansen-Jacobs, M.J., *et al.*, 304,
375 (2-3); 438 (1-2); 828, 835,
1017, 1021, 1252, 1863, 2297,
2315, 5699 (2-3)
Jenman, G.S., 32 (2-3); 1718,
2355 (1-1); 3955, 4263 (2-3);
4326(1-1); 5077, 5270 (2-3)
Knapp, S. & J. Mallet, 2759 (2-3)
Maas, P.J.M., *et al.*, 3609 (2-2);
7507, 7566 (2-3)
McDowell, T., 2338 (1-1); 4645
(2-3)
Mori, S.A., *et al.*, 8009 (2-3);
24466 (1-1)
Mutchnick, P., 52 (2-3); 269 (1-
1); 276, 579, 615, 751, 1136,
1500 (2-3)
Parker, C.S., s.n. (1-1)
Pennington, R.T. & I. Johnson,
435 (2-3)
Persaud, A.C., 180 (2-3); 238 (1-
1)
Samuels, J.A., s.n. (1-1)
Sandwith, N.Y., 121 (1-1); 178,
677 (2-3)

Geay, F., s.n. (2-3)
Granville, J.J. de, *et al.*, 166, 431, 3143, 4163 (2-3); 4798 (2-2); 4811 (2-1); 4857, 5190, 5245, 5252, 5290 (2-3); 5660, 6045, 6298 (2-2); 7190, 7461, 7890, 10567 (2-3); 11710 (1-1); 12623, 12652, 12780 (2-3)
Hallé, F., 1028 (1-1)
Herb. Vahl, **s.n.** (1-2)
Hladik, A., 3095 (1-1)
Irwin, H.S., *et al.*, 48431 (2-3)
Leblond, J.B., 347, 454, s.n. (1-1); s.n. (2-3)
Leeuwenberg, A.J.M., 11638 (1-1)
Lemée, A., s.n. (1-1)
Le Moult, E., 14 (2-3)
Leprieur, F.R., 256, 278 (2-3); s.n., s.n. (1838), s.n. (1840), s.n. (1849), s.n. (1850) (1-1)
Loubry, D., 828 (2-3)
Martin, J., 92, **s.n.** (1-2); s.n. (1-1); s.n. (1842) (2-3); s.n. (1845) (2-3)
Mélinon, M., 113, 137, 139 (2-3); 171 (2-1); 182, 218, 225, 229 (2-3); 233 (1-2); 284, 319, 326 (2-3); 332, 393 (2-1); 406, 433 (2-3); s.n. (1-1); s.n. (2-1)
Monnier, L., s.n. (2-3)
Monot, M., 52 (2-2)
Mori, S.A., *et al.*, 8716, 14829, 15021, 15058, 18058, 18082 (2-3); 18119 (2-1); 18135 (1-1); 22685 (2-3)
Oldeman, R.A.A., 1056 (1-1); 2192, T894 (2-3); B2444 (2-4); 3048 (1-2)
Patris, J.B., **s.n.** (1-1)
Perrottet, G.S., s.n. (1-1); s.n. (2-3)
Poiteau, P.A., s.n. (1-1); s.n. (2-3)

Prévost, M.F., 323, 1083, 1364, 1617 (2-3); 1981 (2-4); 2976 (2-3); 2189 (2-1); 2530, 2571 (2-4); 2964 (2-3)
Richard, L.C., s.n. (1-2); **s.n.** (1-1); s.n. (2-3)
Rothery, 117 (1-1)
Sabatier, D., *et al.*, 2020, 2053, 2441 (2-2); 2530 (2-4); 2863 (2-2); 3177 (2-3); 3141 (2-4); 3194 (2-1)
Sagot, P., 191 (1-2); 192 (2-3); 1275, s.n. (2-1)
Sastre, C., 5502 (2-2); 5950 (1-1)
Sauvain, M., 155 (2-3)
Service Forestier, 242M (2-3)
Skog, L. & C. Feuillet, 7168, 7290 (2-3)
Wachenheim, H., 15, 210 (2-3); 221 (2-1); 250, 254 (1-1); 281, 292 (2-1); 356 (2-3); 408 (2-1); 438, 498 (2-3); s.n. (1-1); s.n. (2-3)
Weitzman, A., *et al.*, 321 (2-3)
without collector, **s.n.** (2-3)

Limnocharitaceae

GUYANA

Harrison, S.G., *et al.*, 1613, 1663 (1-1)
Hekking, W.H.A., 1241c (1-1)
Hitchcock, A.S., 16619 (1-1)
Horn, C.N., *et al.*, 4542, 10003 (1-1)
Irwin, H.S., 285 (1-1)
Jenman, J.S., 5273, 5750 (1-1)
Persaud, A.C., 377 (1-1)
Potter, V.D., 5414 (1-1)
Schomburgk, Ro., ser. I, 562 (1-1)

SURINAME

Cramer, J., LBB 14912 (1-1)
Jonker, F.P., *et al.*, 540 (1-1)
Lanjouw, J. & J.C. Lindeman,
1152, 3154, 3194 (1-1)
Maguire, B. & G. Stahel, 23590
(1-1)
Meulen, J.G.J. van der, L.O.18
(1-1)
Reijenga, T.W., 62 (1-1)
Teunissen, P., LBB 14470 (1-1)
Tulleken, J.E., 278 (1-1)

Alismataceae

GUYANA

Andel, T.R. van, *et al.*, 1107 (1-3)
Appun, C.F., 2164 (1-7b)
Clarke, H.D., 2469 (2-1)
Collector unknown, 388 (1-7a)
Cook, C.D.K., 21 (1-7a); 81 (2-
3); 207a (1-1)
Cooper, A., 436 (2-2)
Cremers, G., 5282 (2-2)
Cruz, J.S. de la, 1272 (2-2); 4097
(2-2)
Forest Dept., G397 (1-1)
Gillespie, L.J., *et al.*, 1186 (2-2);
2006 (2-1)
Goodland, R., 812 (1-4); 977 (2-3)
Görts-van Rijn, A.R.A., *et al.*,
242(1-8)
Graham, E.H., 389 (1-8)
Grewal, M.S. & R. Persaud, 53
(2-2)
Harrison, S.G. *et al.*, 1562 (2-2);
1727 (2-2)
Hekking, W.H.A., 1253 (2-2)
Hitchcock, A.S., 16887 (2-2);
16997 (2-3); 17044 (2-1);
17107 (2-2)

Horn, C.N., *et al.*, 4537 (2-1);
10005 (2-1); 10055 (2-1);
10093 (2-3); 10101 (2-3);
10102 (2-1); 10103 (1-4);
11028 (2-3); 11044 (2-3);
11072 (1-7a); 11079 (2-1);
11083 (1-8)
Jansen-Jacobs, M.J., *et al.*, 70 (1-
7a); 71 (1-5); 486 (2-1); 2128
(1-1); 2592 (2-1); 2623 (2-3);
3214 (1-8); 3628 (2-3); 4601
(2-1); 4999 (2-3); 5037 (1-4);
5082 (1-1); 5084 (1-2)
Jenman, G.S., 1079 (1-7a); **4310**
(1-4); 4656 (2-2); 4992 (2-2);
5080 (1-4); 5162(1-4); 5272
(1-4); 6296 (2-2)
Linder, D.H., 148 (2-1)
Maas, P.J.M., *et al.*, 3768 (2-3);
4072 (1-7a); 7691 (2-1); 7718
(2-1)
Maguire, B. & D.B. Fanshawe,
23568 (2-2)
Persaud, A.C., 354 (2-2)
Persaud, R. & M.S. Grewal, 18
(2-2)
Petter, D., 5365 (2-1)
Rudge, E., s.n. (2-3)
Sandwith, N.Y., 1023 (2-1);
1550 (2-2)
Schomburgk, Ro., add. ser.
163.S (1-1); add. ser. **220.S**
(1-5); ser. I, 563 (1-7a); ser. I,
756 (2-3)
Smith, A.C., 2285 (1-1); 2289
(2-1)
Stoffers, A.L. & A.R.A. Görts-
van Rijn *et al.*, 425 (1-4); 535
(1-7a); 538 (2-3)
Tate, G.H.H., 9 (2-3)
Thurn, E.F. im, s.n., (1-7a)
Tutin, T.G., 95 (2-2)

INDEX TO SYNONYMS, NAMES IN NOTES AND SOME TYPES

Cyrillaceae

Clethra L., see family, note
Clethraceae, see family, note
Cliftonia Banks ex C.F. Gaertn., see family, distribution
Cyrilla
  brevifolia N.E. Br. = 1-1
  racemiflora L. var. *brevifolia* (N.E. Br.) Steyerm. = 1-1
Ericaceae, see family, note
Purdiaea Planch., see family, note; see extra limital taxa
  nutans Planch., see family, distribution; see extra limital taxa

Theophrastaceae

Clavija
  chermontiana Standl. = 1-2b
  macrocarpa Ruiz & Pav., see 1, type
  ornata D. Don var. *subintegra* A. DC. = 1-2a
Theophrasta
  macrophylla Link ex Roem. & Schult. = 1-3

Rhabdodendraceae

Lecostemon
  amazonicum Spruce ex Benth. = 1-1
  crassipes Spruce ex Benth. = 1-1
  crassipes Spruce ex Benth. var. *cayennense* Benth. =1-1
  sylvestre Gleason= 1-1
Rhabdodendron
  arirambae Huber = 1-1
  columnare Gilg & Pilg., see 1, type
  crassipes (Spruce ex Benth.) Huber =1-1
  duckei Huber =1-1
  longifolium Huber =1-1
  macrophyllum (Spruce ex Benth.) Huber, see 1 and 1-1, note
  paniculatum Huber =1-1
  sylvestre (Gleason) Maguire =1-1

Proteaceae

Adenostephanus
    glaziovii Mez = 1-1
    guyanensis Meisn. = 1-2
    sprucei Meisn. = 2-1
Andriapetalum
    cayennense Klotzsch ex Meisn. = 2-2
    rubescens Pohl = 2-1
    rubescens Pohl var. *acuminatum* Meisn. = 2-1
    sessilifolium (Rich.) Klotzsch = 2-2
Euplassa
    meridionalis Knight = 1-2; see 1, type
    venezuelana Steyerm. = 1-1
Panopsis
    acuminata (Meisn.) J.F. Macbr. = 2-1
    cuaensis Steyerm. = 2-1
    hameliifolia (Rudge) Knight = 2-2; see 2, type
Roupala
    angustifolia Diels = 3-3
    chimantensis Steyerm. = 3-4
    complicata Kunth = 3-1
    dentata R. Br. = 3-1
    griotii Steyerm. = 3-5
    hameliifolia Rudge = 2-2; see 2, type
    macropoda Klotzsch & H. Karst. = 3-1
    media R. Br. = 3-1
    montana Aubl. var. *complicata* (Kunth) Griseb. = 3-1
    montana Aubl. var. *dentata* (R. Br.) Sleumer = 3-1
    obtusata Klotzsch var. *obovata* Huber = 3-3
    paruensis Steyerm. = 3-4
    pinnata Lam. = 1-2
    pullei Mennega = 3-5
    pyrifolia Knight = 3-1
    suaveolens Klotzsch var. *minor* Meisn. = 3-5
    schomburgkii Klotzsch = 3-5
    schulzii Mennega = 3-4
    sessilifolia Rich. = 2-2
    yauaperyensis Barb. Rodr. = 2-1

Combretaceae

Avicennia, see 4-1

Buceras
   bucida Crantz, see 5, note
Buchenavia
   capitata (Vahl) Eichler = 1-12; see 1, type
   discolor Diels = 1-8
   huberi Ducke = 1-3
   longibracteata Fróes = 1-1
   macrophylla Eichler, see note 1-4, 1-6
   megalophylla Van Heurck & Müll. Arg., see note 1-4, 1-6
   ptariensis Steyerm. = 1-12
   pulcherrima Exell & Stace = 1-11
   reticulata Eichler, see note 1-4, 1-6
Bucida L. = 5
   angustifolia DC. = 5-1
   buceras L., see 5, type and note
   buceras L. var. *angustifolia* (DC.) Eichler = 5-1
   capitata Vahl = 1-12; see 1, type
Cacoucia Aubl. = 2
   coccinea Aubl. = 2-1; see 2, type
Chuncoa
   amazonia J.F. Gmel. = 5-1
   oblonga (Ruiz & Pav.) Pers. = 5-6
Combretum
   accedens Van Heurck & Müll. Arg. = 2-4
   aubletii DC. = 2-8
   aurantiacum Benth. = 2-2
   brunnescens Gleason = 2-4
   coccineum (Sonn.) Lam., see 2-1
   fulgens Gleason = 2-4
   glabrum DC. = 2-4, see note 2-3
   glaucocarpum Mart., see 2, type
   guianense Miq. = 2-8
   indicum (L.) Jongkind, see 2, note
   laurifolium Mart. = 2-6; see note 2-6
   laxum Aubl., non Jacq., = 2-8
   laxum Jacq., see note 2-3, 2-6
   magnificum Mart. = 2-8
   nitidum Spruce ex Eichler = 2-6; see note 2-6
   obtusifolium Rich. = 2-4
   phaeocarpum Mart. = 2-6
   puberum Rich. = 2-4
   punctatum Steud., non Blume, = 2-8
   rohrii Exell, see note 2-2
   rotundifolium Rich., see 2-2, note 2-2

sprucei Eichler = 2-5
terminalioides Steud. = 2-4
ulei Exell = 2-4
Conocarpus
  lancifolius Engl. & Diels, see distribution 3
  racemosus L. = 4-1; see 4, type
Cordia
  tetraphylla Aubl. = 1-12
Gaura
  fruticosa Loefl. = 2-2; see 2, type
Gimbernatea
  oblonga Ruiz & Pav. = 5-6
Laguncularia
  obovata Miq. = 4-1
Myrobalanus
  catappa (L.) Kuntze = 5-2
  guianensis (Aubl.) Kuntze = 5-3
  lucida (Hoffmanns. ex Mart. & Zucc.) Kuntze = 5-5
  nitidissima (Rich.) Kuntze = 1-7
  oblonga (Ruiz & Pav.) Kuntze = 5-6
Pamea
  guianensis Aubl. = 1-4; see note 1-6
Quisqualis L. = 2
  indica L., see 2, type and note
Rhizophora, see 3-1, 4-1
Tanibouca Aubl. = 5
  guianensis Aubl. = 5-3; see 5, type
Terminalia
  buceras (L.) C. Wright, see 5, type and note
  capitata (Vahl) Sauvalle = 1-12
  dichotoma G. Mey., see 5, type
  eriantha Benth. = 5-5
  guyanensis Eichler, see 5-3
  latifolia Sw. var. *dichotoma* (G. Mey.) DC. = 5-3
  nitidissima Rich. = 1-7
  obidensis Ducke = 5-6
  odontoptera Van Heurck & Müll. Arg. = 5-1
  pamea DC. = 1-4
  paraensis Mart. = 5-2
Thiloa Eichler = 2
  glaucocarpa (Mart.) Eichler, see 2, type
  inundata Ducke = 2-5
  paraguariensis Eichler = 2-5

Dichapetalaceae

Chailletia
  pedunculata DC. = 1-1
  sessiliflora DC. = 2-3
  vestita Benth. = 1-2
Cordia scandens Poir. = 1-2
Dichapetalum
  flavicans Engl. = 1-2
  glabrum Prance = 1-1
  madagascariense Poir., see 1, type
  scandens (Poir.) I.M. Johnst. = 1-2
  vestitum (Benth.) Baill. = 1-2
  vestitum (Benth.) Baill. var. *scandens* Benth. ex Baill. = 1-2
Stephanopodium Poepp., see key to genera
Symphyllanthus
  glaber Vahl = 1-1
  rugosus Vahl = 1-2
Tapura
  amazonica Poepp. var. *ciliata* (Gardn.) Baill. = 2-1
  amazonica Poepp. var. *cuspidata* Baill. = 2-1
  amazonica Poepp. var. *dasyphylla* Baill. = 2-1
  amazonica Poepp. var. *sublanceolata* Baill. = 2-1
  ciliata Gardn. = 2-1
  cucullata Benth. = 2-3
  negrensis Suess. = 2-3

Limnocharitaceae

Hydrocleys
  commersonii Rich., see 1, type
Limnocharis
  flava (L.) Buchenau, see family, note
Stratiotes
  nymphoides Willd. = 1-1

Alismataceae

Alisma
  andrieuxii Hook. & Arn. = 1-7b
  bolivianum Rusby = 1-1
  echinocarpum (Mart.) Seub. = 2-1

intermedium Mart. = 1-7a
macrophyllum Kunth = 1-4
rostratum Nutt., see 1, type
subalatum Mart. = 1-7a
tenellum Mart. = 1-8
Echinodorus
    andrieuxii (Hook. & Arn.) Small = 1-7b
    berteroi (Spreng.) Fassett, see 1, note
    grandiflorus (Cham. & Schltdl.) Micheli var. *floribundus* (Seub.)
        Micheli, see 1-4, note
    grandiflorus (Cham. & Schltdl.) Micheli subsp. *aureus* (Fassett)
        R.R. Haynes & Holm-Niels., see 1-4, note
    guayanensis (Kunth) Griseb. = 2-1
    intermedius (Mart.) Griseb. = 1-7a; see 1-7a, note
    macrophyllus (Kunth) Micheli subsp. macrophyllus, see 1-4, note
    rostratus (Nutt.) Engelm., see 1, type
    scaber Rataj = 1-4
Lophiocarpus
    guayanensis (Kunth) Micheli = 2-1
Lophotocarpus T. Durand = 2; see 2, note
    guayanensis (Kunth) J.G. Sm. = 2-1
    guayanensis (Kunth) J.G. Sm. var. *echinocarpa* (Mart.) Buchenau = 2-1
Sagittaria
    angustifolia Lindl. = 2-2
    echinocarpa Mart. = 2-1
    lancifolia L. var. *angustifolia* (Lindl.) Griseb. = 2-2
    pugioniformis L. = 2-2
    sagittifolia L., see 2, type
    guayanensis Kunth subsp. lappula (D. Don) Bogin, see 2-1, note

INDEX TO VERNACULAR NAMES

Dichapetalaceae

Limnocharitaceae

Alismataceae

Grosse Amazonas Schwertpflanze
  1-5
Hahnenfussähnlicher Igelschlauch
  1-1
Hahnenfuss Froschlöffel 1-1
Horizontal Amazon plant 1-3
Horizontale Amazonpflanze 1-3
Kleinblättriger Froschlöffel 1-8
Lance-leaved Arrowhead 2-2
Langgriffliger Froschlöffel 1-7a

Lanzett blättriges Pfeilkraut 2-2
Large Amazon Swordplant 1-5
Longstyled Toadspoon 1-7a
Pigmy chain-sword 1-8
Smalblättrige Amazonas
  Schwertpflanze 1-2
Small Leaved Amazon
  Swordplant 1-2
Zartblättriger Froschlöffel 1-8

210

Alphabetic list of families of series A occurring in the Guianas

Defined as in Cronquist, 1981, and numbered in his sequence, with alternative names.
Those published, with chronological fascicle number and year.

Abolbodaceae			Cabombaceae	013		
(see Xyridaceae	182)	15. 1994	Cactaceae	031	18. 1997
Acanthaceae	156	23. 2006	Caesalpiniaceae	088	p.p. 7. 1989
(incl. Thunbergiaceae)			Callitrichaceae	150	
(excl. Mendonciaceae	159)		Campanulaceae	162	
Achatocarpaceae	028	22. 2003	(incl. Lobeliaceae)		
Agavaceae	202		Cannaceae	195	1. 1985
Aizoaceae	030	22. 2003	Canellaceae	004	
(excl. Molluginaceae	036)	22. 2003	Capparaceae	067	
Alismataceae	168	27. 2009	Caprifoliaceae	164	
Amaranthaceae	033	22. 2003	Caricaceae	063	
Amaryllidaceae			Caryocaraceae	042	
(see Liliaceae	199)		Caryophyllaceae	037	22. 2003
Anacardiaceae	129	19. 1997	Casuarinaceae	026	11. 1992
Anisophylleaceae	082		Cecropiaceae	022	11. 1992
Annonaceae	002		Celastraceae	109	
Apiaceae	137		Ceratophyllaceae	014	
Apocynaceae	140		Chenopodiaceae	032	22. 2003
Aquifoliaceae	111		Chloranthaceae	008	24. 2007
Araceae	178		Chrysobalanaceae	085	2. 1986
Araliaceae	136		Clethraceae	072	
Arecaceae	175		Clusiaceae	047	
Aristolochiaceae	010	20. 1998	(incl. Hypericaceae)		
Asclepiadaceae	141		Cochlospermaceae		
Asteraceae	166		(see Bixaceae	059)	
Avicenniaceae			Combretaceae	100	27. 2009
(see Verbenaceae	148)	4. 1988	Commelinaceae	180	
Balanophoraceae	107	14. 1993	Compositae		
Basellaceae	035	22. 2003	(= Asteraceae	166)	
Bataceae	070		Connaraceae	081	
Begoniaceae	065		Convolvulaceae	143	
Berberidaceae	016		(excl. Cuscutaceae	144)	
Bignoniaceae	158		Costaceae	194	1. 1985
Bixaceae	059		Crassulaceae	083	
(incl. Cochlospermaceae)			Cruciferae		
Bombacaceae	051		(= Brassicaceae	068)	
Bonnetiaceae			Cucurbitaceae	064	
(see Theaceae	043)		Cunoniaceae	081a	
Boraginaceae	147		Cuscutaceae	144	
Brassicaceae	068		Cycadaceae	208	9. 1991
Bromeliaceae	189	p.p. 3. 1987	Cyclanthaceae	176	
Burmanniaceae	206	6. 1989	Cyperaceae	186	
Burseraceae	128		Cyrillaceae	071	27. 2009
Butomaceae			Dichapetalaceae	113	27. 2009
(see Limnocharitaceae	167)	27. 2009	Dilleniaceae	040	
Buxaceae	115a		Dioscoreaceae	205	
Byttneriaceae			Dipterocarpaceae	041a	17. 1995
(see Sterculiaceae	050)		Droseraceae	055	22. 2003

Ebenaceae	075	
Elaeocarpaceae	048	
Elatinaceae	046	
Eremolepidaceae	105a	25. 2007
Ericaceae	073	
Eriocaulaceae	184	
Erythroxylaceae	118	
Euphorbiaceae	115	
Euphroniaceae	123a	21. 1998
Fabaceae	089	
Flacourtiaceae	056	
(excl. Lacistemaceae	057)	
(excl. Peridiscaceae	058)	
Gentianaceae	139	
Gesneriaceae	155	26. 2008
Gnetaceae	209	9. 1991
Gramineae		
(= Poaceae	187)	8. 1990
Gunneraceae	093	
Guttiferae		
(= Clusiaceae	047)	
Haemodoraceae	198	15. 1994
Haloragaceae	092	
Heliconiaceae	191	1. 1985
Henriquesiaceae		
(see Rubiaceae	163)	
Hernandiaceae	007	24. 2007
Hippocrateaceae	110	16. 1994
Humiriaceae	119	
Hydrocharitaceae	169	
Hydrophyllaceae	146	
Icacinaceae	112	16. 1994
Hypericaceae		
(see Clusiaceae	047)	
Iridaceae	200	
Ixonanthaceae	120	
Juglandaceae	024	
Juncaginaceae	170	
Krameriaceae	126	21. 1998
Labiatae		
(= Lamiaceae	149)	
Lacistemaceae	057	
Lamiaceae	149	
Lauraceae	006	
Lecythidaceae	053	12. 1993
Leguminosae		
(= Mimosaceae	087)	
+ Caesalpiniaceae	088)	p.p. 7. 1989
+ Fabaceae	089)	
Lemnaceae	179	
Lentibulariaceae	160	
Lepidobotryaceae	134a	
Liliaceae	199	
(incl. Amaryllidaceae)		
(excl. Agavaceae	202)	
(excl. Smilacaceae	204)	

Limnocharitaceae	167	27. 2009
(incl. Butomaceae)		
Linaceae	121	
Lissocarpaceae	077	
Loasaceae	066	
Lobeliaceae		
(see Campanulaceae	162)	
Loganiaceae	138	
Loranthaceae	105b	25. 2007
Lythraceae	094	
Malpighiaceae	122	
Malvaceae	052	
Marantaceae	196	
Marcgraviaceae	044	
Martyniaceae		
Mayacaceae	183	
Melastomataceae	099	13. 1993
Meliaceae	131	
Mendonciaceae	159	23. 2006
Menispermaceae	017	
Menyanthaceae	145	
Mimosaceae	087	
Molluginaceae	036	22. 2003
Monimiaceae	005	
Moraceae	021	11. 1992
Moringaceae	069	
Musaceae	192	1. 1985
(excl. Strelitziaceae	190)	
(excl. Heliconiaceae	191)	
Myoporaceae	154	
Myricaceae	025	
Myristicaceae	003	
Myrsinaceae	080	
Myrtaceae	096	
Najadaceae	173	
Nelumbonaceae	011	
Nyctaginaceae	029	22. 2003
Nymphaeaceae	012	
(excl. Nelumbonaceae	010)	
(excl. Cabombaceae	013)	
Ochnaceae	041	
Olacaceae	102	14. 1993
Oleaceae	152	
Onagraceae	098	10. 1991
Opiliaceae	103	14. 1993
Orchidaceae	207	
Oxalidaceae	134	
Palmae		
(= Arecaceae	175)	
Pandanaceae	177	
Papaveraceae	019	
Papilionaceae		
(= Fabaceae	089)	
Passifloraceae	062	
Pedaliaceae	157	
(incl. Martyniaceae)		

Peridiscaceae	058	
Phytolaccaceae	027	22. 2003
Pinaceae	210	9. 1991
Piperaceae	009	24. 2007
Plantaginaceae	151	
Plumbaginaceae	039	
Poaceae	187	8. 1990
Podocarpaceae	211	9. 1991
Podostemaceae	091	
Polygalaceae	125	
Polygonaceae	038	
Pontederiaceae	197	15. 1994
Portulacaceae	034	22. 2003
Potamogetonaceae	171	
Proteaceae	090	27. 2009
Punicaceae	097	
Quiinaceae	045	
Rafflesiaceae	108	
Ranunculaceae	015	
Rapateaceae	181	
Rhabdodendraceae	086	27. 2009
Rhamnaceae	116	
Rhizophoraceae	101	
Rosaceae	084	
Rubiaceae	163	
(incl. Henriquesiaceae)		
Ruppiaceae	172	
Rutaceae	132	
Sabiaceae	018	
Santalaceae	104	
Sapindaceae	127	
Sapotaceae	074	
Sarraceniaceae	054	22. 2003
Scrophulariaceae	153	
Simaroubaceae	130	
Smilacaceae	204	
Solanaceae	142	
Sphenocleaceae	161	
Sterculiaceae	050	
(incl. Byttneriaceae)		

Strelitziaceae	190	1. 1985
Styracaceae	076	
Suraniaceae	086a	
Symplocaceae	078	
Taccaceae	203	
Tepuianthaceae	114	
Theaceae	043	
(incl. Bonnetiaceae)		
Theophrastaceae	079	27. 2009
Thunbergiaceae		
(see Acanthaceae	156)	23. 2006
Thurniaceae	185	
Thymeleaceae	095	
Tiliaceae	049	17. 1995
Trigoniaceae	124	21. 1998
Triuridaceae	174	5. 1989
Tropaeolaceae	135	
Turneraceae	061	
Typhaceae	188	
Ulmaceae	020	11. 1992
Umbelliferae		
Urticaceae	023	11. 1992
Valerianaceae	165	
Velloziaceae	201	
Verbenaceae	148	4. 1988
(incl. Avicenniaceae)		
Violaceae	060	
Viscaceae	106	25. 2007
Vitaceae	117	
Vochysiaceae	123	21. 1998
Winteraceae	001	
Xyridaceae	182	15. 1994
(incl. Albolbodaceae)		
Zamiaceae	208a	9. 1991
Zingiberaceae	193	1. 1985
(excl. Costaceae	194)	
Zygophyllaceae	133	